The Boy Who Followed
His Father into Auschwitz

The Boy Who Followed His Father into Auschwitz

A True Story of Family and Survival

JEREMY DRONFIELD

HARPER

NEW YORK · LONDON · TORONTO · SYDNEY

HARPER

P.S.™ is a registered trademark of HarperCollins Publishers.

Originally published, in slightly different form, as *The Stone Crusher* in 2018 by Chicago Review Press.

First published in the United Kingdom in 2019 by Penguin Random House UK.

HarperCollins books may be purchased for educational, business, or sales promotional use. For information, please email the Special Markets Department at SPsales@harpercollins.com.

FIRST HARPER PAPERBACKS EDITION PUBLISHED 2020.

Frontispiece (as well as in the P.S.) courtesy of Peter Patten.

Library of Congress Cataloging-in-Publication Data has been applied for.

ISBN 978-0-06-301929-4 (pbk.)
ISBN 978-0-06-301931-7 (library edition)

20 21 22 23 24 lsc 10 9 8 7 6 5 4 3 2 1

To Kurt

and in memory of

Gustav
Tini
Edith
Herta
Fritz

The witness has forced himself to testify. For the youth of today, for the children who will be born tomorrow. He does not want his past to become their future.

—Elie Wiesel, *Night*

Contents

CONTENTS

Preface

This is a true story. Every person in it, every event, twist and incredible coincidence, is taken from historical sources. One wishes that it were not true, that it had never occurred, so terrible and painful are some of its events. But it all happened, within the memory of the still living.

There are many Holocaust stories, but not like this one. The tale of Gustav and Fritz Kleinmann, father and son, contains elements of all the others but is quite unlike any of them. Very few Jews experienced the Nazi concentration camps from the first mass arrests in the late 1930s through to the Final Solution and eventual liberation. None, to my knowledge, went through the whole inferno together, father and son, from beginning to end, from living under Nazi occupation, to Buchenwald, to Auschwitz and the prisoner resistance against the SS, to the death marches, and then on to Mauthausen, Mittelbau-Dora, Bergen-Belsen—and made it home again alive. Certainly none who left a written record. Luck and courage played a part, but what ultimately kept Gustav and Fritz living was their love and devotion to each other. "The boy is my greatest joy," Gustav wrote in his secret diary in Buchenwald. "We strengthen each other. We are one, inseparable." This tie had its ultimate test a year later, when Gustav was transported to Auschwitz—a near-certain death sentence—and Fritz chose to cast aside his own safety in order to accompany him.

I have brought the story to life with all my heart. It reads like a novel. I am a storyteller as much as a historian, and yet I haven't needed to invent or embellish anything; even the fragments of dialogue are quoted or reconstructed from primary sources. The bedrock is the concentration camp diary written by Gustav Kleinmann between October 1939 and July 1945, supplemented by a memoir and interviews given by Fritz in 1997. None of these sources makes easy reading, either emotionally or literally—the diary, written under extreme circumstances, is sketchy, often making cryptic allusions to things beyond the knowledge of the general reader (even Holocaust historians would have to consult their reference works to interpret some passages). Gustav's motive in writing it was not to make a record but to help preserve his own sanity; its references were comprehensible to him at the time. Once unlocked, it provides a rich and harrowing insight into living the Holocaust week by week, month by month, and year after year. Strikingly, it reveals Gustav's unbeatable strength and spirit of optimism: ". . . every day I say a prayer to myself," he wrote in the sixth year of his incarceration: *Do not despair. Grit your teeth—the SS murderers must not beat you.*"

Interviews with surviving members of the family have provided additional personal detail. The whole—from Vienna life in the 1930s to the functioning of the camps and the personalities involved—has been backed up by extensive documentary research, including survivor testimony, camp records, and other official documents, which have verified the story at every step of the way, even the most extraordinary and incredible.

Jeremy Dronfield, June 2018

The Boy Who Followed
His Father into Auschwitz

Prologue

Fritz Kleinmann shifted with the motion of the train, shuddering convulsively in the subzero gale roaring over the sidewalls of the open freight wagon. Huddled beside him, his father dozed, exhausted. Around them sat dim figures, moonlight picking out the pale stripes of their uniforms and the bones in their faces. It was time for Fritz to make his escape; soon it would be too late.

Eight days had passed since they'd left Auschwitz on this journey. They had walked the first forty miles, the SS driving the thousands of prisoners westward through the snow, away from the advancing Red Army. Intermittent gunshots were heard from the rear of the column as those who couldn't keep up were murdered. Nobody looked back.

Then they'd been put on trains bound for camps deeper inside the Reich. Fritz and his father managed to stay together, as they had always done. Their transport was for Mauthausen in Austria, where the SS would carry on the task of draining the last dregs of labor from the prisoners before finally exterminating them. One hundred and forty men crammed into each open-topped wagon—at first they'd had to stand, but as the days passed and the cold killed them off, it gradually became possible to sit down. The corpses were stacked at one end of the wagon and their clothing taken to warm the living.

They might be on the brink of death, but these prisoners were the lucky ones, the useful workers—most of their brothers and sisters, wives, mothers, and children had been murdered or were being force-marched westward and dying in droves.

Fritz had been a boy when the nightmare began seven years ago; he'd grown to manhood in the Nazi camps, learning, maturing, resisting the pressure to give up hope. He had foreseen this day and prepared for it. Beneath their camp uniforms he and his papa wore civilian clothing, which Fritz had obtained through his friends in the Auschwitz resistance.

The train had passed through Vienna, the city that had once been their home, then turned west, and now they were only ten miles from their destination. They were back in their homeland, and once they broke free they could pass for local workmen.

Fritz had been delaying the moment, worried about his father. Gustav was fifty-three years old and exhausted—it was a miracle he had survived this far. Now that it came to it, he could not make the escape attempt. The strength wasn't in him anymore. Yet he couldn't deny his son the chance to live. It would be a wrenching pain to part after so many years of helping one another to survive, but he urged Fritz to go alone. Fritz begged him to come, but it was no good: "God protect you," his father said. "I can't go, I'm too weak."

If Fritz didn't make the attempt soon, it would be too late. He stood up and changed out of the hated uniform; then he embraced his papa, kissed him, and with his help climbed the slippery sidewall of the wagon.

The full blast of the wind at minus four degrees hit him hard. He peered anxiously toward the brake houses on the adjacent wagons, occupied by armed SS guards. The moon was bright—two days from the full, rising high and laying a ghostly glow across the snowy landscape, against which any moving shape would be starkly visible.[1] The train was thundering along at its maximum speed. Screwing up his courage and hoping for the best, Fritz launched himself into the night and the rushing, freezing air.

PART I
Vienna
Seven Years Earlier . . .

1. "When Jewish Blood Drips from the Knife . . ."

אבא

Gustav Kleinmann's lean fingers pushed the fabric under the foot of the sewing machine; the needle chattered, machine-gunning the thread into the material in a long, immaculate curve. Next to his worktable stood the armchair it was intended for, a skeleton of beechwood with taut webbing sinews and innards of horsehair. When the panel was stitched, Gustav fitted it over the arm; his little hammer drove in the nails—plain tacks for the interior, studs with round brass heads for the outer edge, tightly spaced like a row of soldiers' helmets; in they went with a *tap-tapatap*.

It was good to work. There wasn't always enough to go around, and life could be precarious for a middle-aged man with a wife and four children. Gustav was a gifted craftsman but not an astute businessman, although he always muddled through. Born in a tiny village by a lake in the historic kingdom of Galicia,* a province of the Austro-Hungarian Empire, he'd come to Vienna at age fifteen to train as an upholsterer, and then settled here. Called to military service in the spring of the year he turned twenty-one, he'd served in the Great War, been wounded twice and decorated for bravery, and at the war's end he'd returned to Vienna to resume his humble trade, working his way up to master craftsman. He had married his girl, Tini, during the war, and together they had raised four fine, happy children. And there was Gustav's life: modest,

* Now part of southern Poland and western Ukraine.

hardworking; and if not entirely content, he was at least in-clined to be cheerful.

The droning of airplanes interrupted Gustav's thoughts; it grew and receded as if they were circling over the city. Curi-ous, he laid down his tools and stepped out into the street.

Im Werd was a busy thoroughfare, noisy with the clop and clatter of horse-drawn carts and the grumbling of trucks, the air thick with the smells of humanity, fumes, and horse dung. For a confusing moment it appeared to Gustav to be snowing—in March!—but it was a blizzard of paper fluttering from the sky, settling on the cobbles and the market stalls of the Kar-melitermarkt. He picked one up.

PEOPLE OF AUSTRIA!

For the first time in the history of our Fatherland, the leadership of the state requires an open commitment to our Homeland . . .[1]

Propaganda for this Sunday's vote. The whole country was talking about it, and the whole world was watching. For every man, woman, and child in Austria it was a big deal, but for Gustav, as a Jew, it was of the utmost importance—a national vote to settle whether Austria should remain independent from German tyranny.

For five years, Nazi Germany had been looking hungrily across the border at its Austrian neighbor. Adolf Hitler, an Austrian by birth, was obsessed with the idea of bringing his homeland into the German Reich. Although Austria had its own homegrown Nazis eager for unification, most Austrians were opposed to it. Chancellor Kurt Schuschnigg was under pressure to give members of the Nazi Party positions in his government, Hitler threatening dire consequences if he didn't comply—Schuschnigg would be forced out of office and replaced with a Nazi puppet; unification would follow, and Austria would be swallowed by Germany. The country's 183,000 Jews regarded this prospect with dread.[2]

The world watched keenly for the outcome. In a desperate last throw of the dice, Schuschnigg had announced a plebiscite—a referendum—in which the people of Austria would decide for themselves whether they wanted to keep their independence. It was a courageous move; Schuschnigg's predecessor had been assassinated during a failed Nazi coup, and right now Hitler was ready to do just about anything to prevent the vote going ahead. The date had been set for Sunday, March 13, 1938.

Nationalist slogans ("Yes for Independence!") were pasted and painted on every wall and pavement. And today, with two days to go until the vote, planes were showering Vienna with Schuschnigg's propaganda. Gustav looked again at the leaflet.

> . . . For a free and Germanic, independent and social, Christian and united Austria! For peace and work and equal rights for all who profess allegiance to the people and the Fatherland.
>
> . . . The world shall see our will to live; therefore, people of Austria, stand up as one man and vote YES![3]

These stirring words held mixed meanings for the Jews. They had their own ideas of Germanism—Gustav, immensely proud of his service to his country in the Great War, considered himself an Austrian first and a Jew second.[4] Yet he was excluded from Schuschnigg's Germanic Christian ideal. He also had reservations about Schuschnigg's Austrofascist government. Gustav had once been an organizer for the Social Democratic Party of Austria. With the rise of the Austrofascists in 1934, the party had been violently suppressed and outlawed (along with the Nazi Party).

But for the Jews of Austria at this moment, anything was preferable to the kind of open persecution going on in

Germany. The Jewish newspaper *Die Stimme* had a banner in today's edition: "We support Austria! Everyone to the ballot boxes!"[5] The Orthodox paper *Jüdische Presse* made the same call: "No special request is needed for the Jews of Austria to come out and vote in full strength. They know what this means. Everyone must fulfil his duty!"[6]

Through secret channels, Hitler had threatened Schuschnigg that if he didn't call off the plebiscite, Germany would take action to prevent it. At this very moment, while Gustav stood in the street reading the leaflet, German troops were already massing at the border.

אמא

With a glance in the mirror, Tini Kleinmann patted down her coat, gathered her shopping bag and purse, left the apartment, and woke the echoes in the stairwell with her neat little heels *click-clacking* briskly down the flights. She found Gustav standing in the street outside his workshop, which was on the ground floor of the apartment building. He had a leaflet in his hand; the road was littered with them—in the trees, on the rooftops, everywhere. She glanced at it and shivered; Tini had a feeling of foreboding about it all which Gustav the optimist didn't quite share. He always thought things would work out for the best; it was both his weakness and his strength.

Tini walked briskly across the cobbles to the market. A lot of the stallholders were peasant farmers who came each morning to sell their produce alongside the Viennese traders. Many of the latter were Jews; indeed, more than half the city's businesses were Jewish owned, especially in this area. Local Nazis capitalized on this fact to stir up anti-Semitism among the workers suffering in the economic depression—as if the Jews were not suffering from it too.

Gustav and Tini weren't particularly religious, going to synagogue perhaps a couple of times a year for anniversaries and memorials, and like most Viennese Jews, their children bore Germanic rather than Hebraic names, yet they followed the Yiddish customs like everyone else. From Herr Zeisel the butcher Tini bought veal, thinly sliced for Wiener schnitzel; she had leftover chicken for the Shabbat* evening soup, and from the farm stalls she bought fresh potatoes and salad; then bread, flour, eggs, butter . . . Tini progressed through the bustling Karmelitermarkt, her bag growing heavier. Where the marketplace met Leopoldsgasse, the main street, she noticed the unemployed cleaning women touting for work; they stood outside the Klabouch boarding house and the coffee shop. The lucky ones would be picked up by well-off ladies from the surrounding streets. Those who brought their own pails of soapy water got the full wage of one schilling.† Tini and Gustav sometimes struggled to pay their bills, but at least she hadn't been reduced to *that*.

The proindependence slogans were everywhere, painted on the pavements in big, bold letters like road markings: the rallying cry for the plebiscite—"We say yes!"—and everywhere the Austrian "crutch cross."‡ From open windows came the sound of radios turned up high, playing cheerful patriotic music. As Tini watched, there was a burst of cheering and a roar of engines as a convoy of trucks came down the street, filled with uniformed teenagers of the Austrian Youth waving banners in the red-and-white national colors and flinging out more

---

* Sabbath; from just before sundown on Friday to darkness on Saturday evening.
† Equivalent to about two or three dollars in 2019.
‡ A cross with T-bars at the ends of the arms.

leaflets.[7] Bystanders greeted them with fluttering handker-chiefs, doffed hats, and cries of "Austria! Austria!"

It looked as if independence was winning . . . so long as you took no notice of the sullen faces among the crowds. The Nazi sympathizers. They were exceptionally quiet today—and exceptionally few in number, which was strange.

Suddenly the cheerful music was interrupted and the radios crackled with an urgent announcement—all unmarried army reservists were to report immediately for duty. The purpose, said the announcer, was to ensure order for Sunday's plebiscite, but his tone was ominous. Why would they need extra troops for that?

Tini turned away and walked back through the crowded market, heading for home. No matter what occurred in the world, no matter how near danger might be, life went on, and what could one do but live it?

בן

Across the city the leaflets lay on the waters of the Danube Canal, in the parks and streets. Late that afternoon, when Fritz Kleinmann left the Trade School on Hütteldorfer Strasse on the western edge of Vienna, they were lying in the road and hanging in the trees. Roaring down the street came column after column of trucks filled with soldiers, heading for the German border one hundred and seventy miles away. Fritz and the other boys watched excitedly, as boys will, as rows of helmeted heads sped past, weapons ready.

At fourteen years old, Fritz already resembled his father—the same handsome cheekbones, the same nose, the same mouth with full lips curving like a gull's wings. But whereas Gustav's countenance was gentle, Fritz's large, dark eyes were penetrating, like his mother's. He'd left high school, and for the past six months had been training to enter his father's trade as an upholsterer.

As Fritz and his friends made their way homeward through the city center, a new mood was taking hold of the streets. At three o'clock that afternoon the government's campaigning for the plebiscite had been suspended due to the developing crisis. There was no official news, only rumors: of fighting on the Austrian-German border; of Nazi uprisings in the provincial towns; and, most worrying of all, a rumor that the Viennese police would side with local Nazis if it came to a confrontation. Bands of enthusiastic men had begun roaming the streets—some yelling "Heil Hitler!" and others replying defiantly "Heil Schuschnigg!" The Nazis were louder, growing bolder, and most of them were youths, empty of life experience and pumped full of ideology.[8]

This sort of thing had been going on sporadically for days, and there had been occasional violent incidents against Jews;[9] but this was different—when Fritz reached Stephansplatz, right in the very heart of the city, where Vienna's Nazis had their secret headquarters, the space in front of the cathedral was teeming with yelling, baying people; here it was all "Heil Hitler" and no counterchant.[10] Policemen stood nearby, watching, talking among themselves, but doing nothing. Also watching from the sidelines, not yet revealing themselves, were the secret members of the Austrian Sturmabteilung—the SA, the Nazi Party's storm troopers. They had discipline, and they had their orders; their time hadn't yet arrived.

Avoiding the knots of demonstrators, Fritz crossed the Danube Canal into Leopoldstadt and was soon back in the apartment building, his boots clattering up the stairs to number 16—home, warmth, and family.

משפחה

Little Kurt stood on a stool in the kitchen, watching as his mother prepared the noodle batter for the chicken soup, the traditional Shabbat Friday meal. It was one of the few traditional practices the family maintained; Tini lit no candles, said no blessing. Kurt was different—only eight years old, he sang in the choir of the city-center synagogue and was becoming quite devout. He'd made friends with an Orthodox family who lived across the hallway, and it was his role to switch on the lights for them on Shabbat evenings.

He was the baby and the beloved; the Kleinmanns were a close family, but Kurt was Tini's particular darling. He loved to help her cook.

While the soup simmered, he watched, lips parted, as she whipped the egg batter to a froth and fried it into thin pancakes. This was one of his favorite cooking duties. The very best was Wiener schnitzel, for which his mother would gently pound the veal slices with a tenderizer until they were as soft and thin as velvet; she taught him to coat them in the dish of flour, the batter of beaten egg and milk, and finally the breadcrumbs; then she would lay them two by two in the pan of bubbling, buttery oil, the rich aroma filling the little apartment as the cutlets puffed and crinkled and turned golden. Tonight, though, it was the smell of fried noodles and chicken.

From the next room—which doubled as bedroom and living room—came the sound of a piano; Kurt's sister Edith, eighteen years old, played well, and had taught Kurt a pleasant little tune called "Cuckoo," which would remain in his memory forever. His other sister, Herta, aged fifteen, he simply adored; she was closer to him in age than Edith, who was a grown woman. Herta's place in Kurt's heart would always be as an image of beauty and love.

Tini smiled at his earnest concentration as he helped her roll up the cooked egg, slicing it into noodles, which she stirred into the soup.

The family sat down to their meal in the warm glow of the Shabbat—Gustav and Tini; Edith and Herta; Fritz and little Kurt. Their home was small—just this room and the bedroom which they all shared (Gustav and Fritz in together, Kurt with their mother, Edith in her own bed, and Herta on the sofa); yet home it was, and they were happy here.

Outside, a shadow was gathering over their world. That afternoon, a written ultimatum had come from Germany, insisting that the plebiscite be canceled; that Chancellor Schuschnigg resign; that he be replaced by the right-wing politician Arthur Seyss-Inquart (a secret member of the Nazi Party) with a sympathetic cabinet under him. Hitler's justification was that Schuschnigg's government was repressing the ordinary Germans of Austria ("German" being synonymous with "Nazi" in Hitler's mind). Finally, the exiled Austrian Legion, a force of thirty thousand Nazis, must be brought back to Vienna to keep order on the streets. The Austrian government had until 7:30 p.m. to comply.[11]

After dinner, Kurt had to hurry off to the Shabbat evening service at the synagogue. He was paid a schilling a time for singing in the choir (substituted by a chocolate bar on Saturday mornings), so it was an economic as well as a religious duty.

As usual, Fritz escorted him; he was an ideal older brother— friend, playmate, and protector. The streets were busy this evening, but the unruly noise had subsided, leaving behind a sense of lurking malevolence. Usually Fritz would accompany Kurt as far as the billiard hall on the other side of the Danube Canal—"You know your way from here, don't you?"—and head off to play billiards with his friends. But this evening that wouldn't do, and they walked together all the way to the Stadttempel.

Back in the apartment, the radio was playing. The program was interrupted by an announcement. The plebiscite had been postponed. It was like an ominous tap on the shoulder. Then, a little after half past seven, the music broadcast was halted, and a voice declared: "Attention! In a few moments you will hear an extremely important announcement." There came a pause, empty, hissing; it lasted for three full minutes, and then Chancellor Schuschnigg came on. His voice wavered with emotion: "Austrian men and Austrian women; this day has placed us in a tragic and decisive situation." Every person in Austria who was near a radio at that moment listened intently, many with fear, some with excitement, as the chancellor described the German ultimatum. Austria must take its orders from Germany or be destroyed. "We have yielded to force," he said, "since we are not prepared even in this terrible situation to shed Germanic blood. We decided to order the troops to offer no serious . . ." He hesitated ". . . to offer no resistance." His voice cracking, he gathered himself for the final words. "So I take my leave of the Austrian people, with a German word of farewell, uttered from the depths of my heart: God protect Austria."[12]

Gustav, Tini, and their daughters sat stunned as the national anthem began to play. In the studio, unseen and unheard by the people, Schuschnigg broke down and sobbed.

בן

The sweet, exalting phrases of the "Hallelujah," led by the cantor's tenor and fleshed out by the voices of the choir, filled the great oval space of the Stadttempel, embracing the marble pillars and the gilded ornamentation of the tiered balconies in harmonious sound. From his place in the choir on the very top

tier behind the Ark,* Kurt could look right down on the *bimah*†
and the congregation. It was far more crowded than usual,
packed to bursting—people driven by uncertainty to seek
comfort in their religion. The religious scholar Dr. Emil
Lehmann, unaware of the latest news, had spoken movingly
about Schuschnigg, exalting the plebiscite, closing with the
now deposed chancellor's rallying cry: "We say yes!"[13]

After the service, Kurt filed down from the balcony, collected his schilling, and found Fritz waiting. Outside, the narrow cobbled lane was thronged with the departing congregation.
From out here there was little to show the synagogue's presence; it appeared to be part of a row of apartment houses—the
main body was behind the façade, squeezed between this street
and the next. While Leopoldstadt was nowadays the Jewish
quarter of Vienna, this little enclave in the old city center,
where Jews had lived since the Middle Ages, was the cultural
heart of Jewish life in Vienna. It was in the buildings and the
street names—Judengasse, Judenplatz—and their blood was
in the cobblestones and in the crevices of history, in the persecutions and the medieval pogrom that had driven them to live
in Leopoldstadt.

By day the narrow Seitenstettengasse was insulated from
much of the noise of the city, but now, in the Shabbat evening
darkness, Vienna was bursting to life. A short distance away, in
the Kärntnerstrasse, a long thoroughfare on the other side of
the Nazi enclave in Stephansplatz, a mob was gathering. The
brown-shirted storm troopers of the SA, free now to bring out
their concealed weapons and put on their swastika armbands,
were on the march. The police marched with them. Trucks

---

* Ornate cabinet in which the scrolls of the Torah are kept.
† Reading table used by a rabbi, facing the Ark.

rolled along filled with storm troopers; men and women danced and yelled by the light of flaming torches.

Across the city came the full-throated roar—"Heil Hitler! *Sieg Heil!* Down with the Jews! Down with the Catholics! One people, one Reich, one Führer, one victory! Down with the Jews!" Raw, fanatical voices rose in song: "Deutschland über Alles" and chanted: "Today we have all Germany—tomorrow we have the world!"[14] The playwright Carl Zuckmayer wrote that "The netherworld had opened its portals and spewed out its basest, most horrid, and filthiest spirits . . . What was being unleashed here was the revolt of envy; malevolence; bitterness; blind, vicious vengefulness."[15] A British journalist who witnessed it called the procession "an indescribable witches' sabbath."[16]

Echoes reached the Seitenstettengasse, where the Jews outside the Stadttempel were dispersing. Fritz shepherded Kurt down the Judengasse and across the bridge. Within minutes they were back in Leopoldstadt.

The Nazis were coming, along with hordes of newfound weathercock friends, flooding in tens of thousands through the city center toward the Jewish district. The tide poured across the bridges into Leopoldstadt, washing into Taborstrasse, Leopoldsgasse, the Karmelitermarkt and Im Werd—a hundred thousand chanting, roaring men and women, filled with triumph and hate. "*Sieg Heil!* Death to the Jews!" The Kleinmanns sat in their home, listening to the tumult outside, waiting for it to burst in through the door.

But it didn't come. For hours the mobs ruled the streets, all noise and fury, but doing little physical harm; some unlucky Jews were caught outdoors and abused; people who "looked Jewish" were beaten up; known Schuschnigg loyalists were attacked; a number of homes and businesses were invaded and plundered, but the storm of destruction did not break over Vienna that night. Amazed, some people wondered whether the

legendarily genteel nature of the Viennese people might temper the behavior even of its Nazis.

It was a vain hope. The reason for the restraint was simple:
the storm troopers were in charge, and they were disciplined,
intending to strip and destroy their prey methodically, not by
riot. Together with the police (now wearing swastika armbands), the SA took over public buildings. Prominent members of the governing party were seized or fled. Schuschnigg
himself was arrested. But this was just a prelude.

By next morning, the first columns of German troops had
crossed the border.

The European powers—Britain, France, Czechoslovakia—
objected to Germany's invasion of sovereign territory, but
Mussolini, supposedly Austria's ally, refused to consider any
military action; he wouldn't even condemn Germany. International resistance fell apart before it had even formed. The world
left Austria to the dogs.

And Austria welcomed them.

אבא

Gustav woke to the sound of engines. A low drone that entered
his skull with the stealth of an odor and grew in volume. Airplanes. For a moment it was as if he were in the street outside
his workshop: it was still yesterday; the nightmare had not happened. It was scarcely breakfast time. The rest of the family,
apart from Tini, clattering quietly in the kitchen, were still in
their beds, just stirring from their dreams.

As Gustav rose and dressed, the droning grew louder. There
was nothing to be seen from the windows—just rooftops and
a strip of sky—so he put on his shoes and went downstairs.

In the street and across the Karmelitermarkt there was little
sign of the night's terrors—just a few stray "Vote Yes!" leaflets,

trampled and swept into corners. The traders were setting out their stalls and opening their shops. Everyone looked to the sky as the rumbling engines grew louder and louder, rattling windows, drowning out the sounds of the streets. This wasn't like yesterday at all—this was an oncoming thunderstorm. The planes came into view over the rooftops. Bombers, dozens of them in tight formation, with fighters darting loose above them. They flew so low that even from the ground their German markings could be picked out and their bomb-bay doors could be seen opening.[17] A ripple of terror swept across the marketplace.

What came out, though, was not bombs but another snowstorm of paper, fluttering down over the roofs and streets. Here was a political climate that produced actual weather. Gustav picked up one of the leaflets. It was briefer and simpler than yesterday's message. At the head was the Nazi eagle, and a declaration:

> National Socialist Germany greets her National Socialist
> Austria and the new National Socialist government.
> Joined in a faithful, unbreakable bond!
> Heil Hitler![18]

The storm of engines was deafening. Not only the bombers but over a hundred transport planes flew over; while the bombers banked and circled, the others headed southeast. Nobody knew it yet, but these were troop-carrying aircraft, heading for Aspern Airfield just outside the city—the first German spearhead into the Austrian capital. Gustav dropped the slip of paper as if it were toxic and went back indoors.

Breakfast was bleak that morning. From this day forward a specter would haunt every move, word, and thought of every Jewish person. They all knew what had happened in Germany in the past five years. What they didn't yet know was that in

Austria there would be no gradual onset; they would experience five years' worth of terror in one frantic torrent.

The Wehrmacht was coming, the SS and Gestapo were coming, and there were rumors that the Führer himself had reached Linz and would soon be in Vienna. The city's Nazis were mad with excitement and triumph. The majority of the populace, wanting only stability and safety, began to sway with the times. Jewish stores in Leopoldstadt were systematically plundered by squads of SA storm troopers, while the homes of wealthier Jews began to be raided and robbed. Envy and hatred against Jews in business, in skilled trades, and in the legal and medical professions had built to a head during the economic depression, and the boil was about to be violently lanced.

There was a myth that it wasn't in the nature of the Viennese to conduct politics through street-fighting and rioting—"The real Viennese," they said in dismay as the Nazis filled the streets with noise and fury, "discusses his differences over a café table and goes like a civilized being to the polls."[19] But in due course "the real Viennese" would go like a civilized being to his doom. The savages ruled this country now.

Yet Gustav Kleinmann, a hopeful man by nature, believed that his family might be safe—they were, after all, Austrians more than Jews. The Nazis would surely only persecute the devout, the openly Hebraic, the Orthodox . . . wouldn't they?

בת

Edith Kleinmann kept her head high as she walked. Like her father she considered herself an Austrian more than a Jew. She thought little of such things—she was eighteen years old; by day she was learning millinery and had ambitions to be a hat designer; in her free hours she had a good time, went out with boys and loved music and dancing. Edith was, above all else, a young woman, with the drives and desires of youth. The boys she went out with were rarely Jewish. This made Gustav uneasy; being Austrian was a fine thing, but he felt that one should still cleave to one's people. If there was a contradiction there, Gustav didn't recognize it.

A few days had passed since the arrival of the Germans. They had marched in on Sunday, the day the abandoned plebiscite would have taken place. Most Jews had stayed indoors, but Edith's brother Fritz, typically daring, had ventured out to watch. At first, he reported, a few brave Viennese threw stones at the German troops, but they were quickly overwhelmed by the cheering, Heil-Hitlering multitude. When the full German force made its triumphal entrance into the capital, led by Adolf Hitler himself, the columns seemed endless: fleets of gleaming limousines, motorcycles, armored cars, thousands of field-gray uniforms, helmets, and tramping jackboots. The scarlet swastika flags were everywhere—held aloft by the soldiers, hanging from the buildings, fluttering from the cars. Behind the scenes, Heinrich Himmler had flown in and begun the process of taking over the police.[20] The plundering of wealthy Jews went on, and suicides were reported daily.

Edith walked briskly. Some kind of disturbance was going on at the corner of the Schiffamtsgasse and Leopoldsgasse, where a large crowd had gathered near the police station.[21] Edith could hear laughter and cheering. She went to cross the road, but slowed her step, noticing a familiar face in the press—Vickerl Ecker, an old schoolfriend. His bright, eager eyes met hers.

"There! She's one!"[22]

Faces turned toward her, she heard the word *Jewess*, and hands gripped her arms, propelling her toward the crowd. She saw Vickerl's brown shirt, the swastika armband. Then she was through the press of bodies and in the midst of a ring of leering, jeering faces. Half a dozen men and women were on their hands and knees with brushes and buckets, scrubbing the pavement—all Jews, all well dressed. One bewildered woman clutched her hat and gloves in one hand and a scrubbing brush in the other, her immaculate coat trailing on the wet stones.

"On your knees." A brush was put in Edith's hand and she was pushed to the ground. Vickerl pointed at the Austrian crosses and *Say Yes!* slogans. "Get rid of your filthy propaganda, Jewess." The spectators crowed as she began to scrub. There were faces she recognized in the crowd—neighbors, acquaintances, smartly dressed businessmen, prim wives, rough workingmen and women, all part of the fabric of Edith's world, transformed into a gloating mob. She scrubbed, but the paint wouldn't come off. "Work suitable for Jews, eh?" somebody called out and there was more laughter. One of the storm troopers picked up a man's bucket and emptied it over him, soaking his camel-hair coat. The crowd cheered.

After an hour or so, the victims were given receipts for their "work" and permitted to go. Edith walked home, stockings torn, clothes soiled, struggling to contain herself, brimming over with shame and degradation.

In the coming weeks these "scrubbing games" became an everyday part of life in Jewish neighborhoods. The patriotic slogans proved impossible to remove, and often the SA added acid to the water so that it burned and blistered the victims' hands.[23] Fortunately for Edith she wasn't taken again, but her fifteen-year-old sister, Herta, was among a group forced to scrub the Austrian crosses from the clock pillar in the marketplace. Other Jews were forced to paint anti-Semitic slogans on Jewish-owned shops and businesses in livid red and yellow.

The suddenness with which genteel Vienna had turned was breathtaking—like tearing the soft, comfortable fabric of a familiar couch to reveal sharp springs and nails beneath. Gustav was wrong; the Kleinmanns were not safe. Nobody was safe.

משפחה

They all dressed in their best outfits before leaving the apartment—Gustav wore his Sunday suit; Fritz in schoolboy knickerbocker trousers; Edith, Herta, and Tini in their smartest dresses; little Kurt in a sailor suit. In Hans Gemperle's photography studio they gazed into the camera's lens as if looking to their own futures. Edith smiled uncomfortably, resting a hand on her mother's shoulder. Kurt looked contented—at eight he understood little of what the changes in his world might mean—and Fritz displayed the nonchalant ease of a cocky teenager, while Herta—just turning sixteen and a young woman already—was radiant. As Herr Gemperle (who was not a Jew and would thrive in the coming years) clicked his shutter, he caught Gustav's apprehensiveness and the stoicism of Tini's dark eyes. They understood now where the world was going, even the sanguine Gustav. It had been Tini's urging that had brought them to the studio. She had a foreboding that the family might not be together for much longer and wanted to capture her children's image while she had the chance.

The poison on the streets now began to flow from the offices of government and justice. Under the Nuremberg Laws of 1935, Austrian Jews were stripped of their citizenship. On April 4, Fritz and all his Jewish school friends were expelled from the Trade School; he also lost his work placement. Edith and Herta were fired from their jobs, and Gustav was no longer able to practice his trade; his workshop was seized and locked up. People were warned not to buy from Jews; those who were caught doing so were made to stand with a sign: "I am an Aryan, but a swine—I bought in this Jewish shop."[24]

Four weeks after the Anschluss,* Adolf Hitler returned to

---

* Literally "joining"; the forcible unification of Austria with Germany.

Vienna. He gave a speech at the Nordwest railway station—only a mile from Im Werd—to a crowd of twenty thousand members of the SA, SS, and Hitler Youth. "I have shown through my life," he thundered, "that I can do more than those dwarfs who ruled this country into ruin. In a hundred years' time my name will stand as that of the great son of this country."[25] The crowd exploded into a storm of "*Sieg Heil!*" repeated over and over, earsplitting, echoing throughout the Jewish neighborhoods of Leopoldstadt.

Vienna was decked with swastikas, every newspaper filled with pictures glorifying the Führer. The next day Austria had its long-awaited plebiscite on independence. Jews, of course, were barred from voting. The ballot was firmly controlled and closely monitored by the SS, and to nobody's surprise the result was 99.7 per cent in favor of the Anschluss. Hitler declared that the result "surpassed all my expectations."[26] The bells of Protestant churches across the city rang for fifteen long minutes, and the head of the Evangelical Church ordered services of thanksgiving. The Catholics remained silent, not yet certain if the Führer meant to deal them Jews' wages.[27]

Foreign newspapers were banned. Swastika lapel badges began to appear everywhere, and suspicion fell on any man or woman not wearing one.[28] In schools, the Heil Hitler salute became part of the daily routine after morning prayers. There were ritual book burnings, and the SS took over the Israelitische Kultusgemeinde, the Jewish cultural and religious affairs center near the Stadttempel, humiliating and baiting the rabbis and other officials who staffed it.[29] From now on the IKG would become the government organ through which the "Jewish problem" was handled, and would have to pay "compensation" to the state to occupy its own premises.[30] The regime seized Jewish property worth a total of two and a quarter billion Reichsmarks (not including houses and apartments).[31]

Gustav and Tini struggled to hold their family together. Gustav had a few good Aryan friends in the upholstery trade who gave him employment in their workshops, but it was infrequent. During the summer, Fritz and his mother got work from the owner of the Lower Austrian Dairy, delivering milk in the neighboring district early in the morning, when the customers wouldn't know that their milk was being brought by Jews. They earned two pfennigs for each bottle they delivered, making up to one mark a day—starvation wages. The family subsisted on meals from the Jewish soup kitchen down the street.

There was no escaping the touch of Nazism. Groups of brown-shirted storm troopers and Hitler Youth marched in the streets singing:

> When Jewish blood drips from the knife,
> Then we sing and laugh.

Their songs extolled the hanging of Jews and putting Catholic priests against the wall. Some of the singers were old friends of Fritz's, who had turned Nazi with shocking suddenness. Some had even joined the local SS unit, the 89th Standarte. The SS were everywhere, demanding identification from passing citizens, proud and pleased in their crisp uniforms and unalloyed power. It infected everything. The word *Saujud*—Jew-pig—was heard everywhere. Signs saying "Aryans only" appeared on park benches. Fritz and his remaining friends were barred from playing on sports grounds or using swimming pools—which struck Fritz hard, because he loved to swim.

As summer progressed, the anti-Semitic violence subsided, but official sanctions went on, and beneath the surface a pressure was building. A fearful name began to be heard: "Keep your head down and your mouth shut," said Jews to one another, "or you will go to Dachau." People began to disappear: prominent figures first—politicians and businessmen—then able-bodied Jewish men were spirited away on flimsy pretexts. Sometimes they were delivered back to their families in ashes. Then another name began to be whispered: Buchenwald. The *Konzentrationslager*—concentration camps—which had been a feature of Nazi Germany since the beginning, were multiplying.[32]

Persecution of Jews was becoming thoroughly bureaucratic. Their identities were a matter of special attention. In August it was decreed that if they didn't already have recognized Hebraic first names, they had to take new middle names—"Israel" for men, "Sara" for women.[33] Their identity cards had to be stamped with a "J"—the *Juden-Kennkarte*, or *J-Karte* as they called it. In Leopoldstadt, a special procedure was employed. The cardholder, having had their card stamped, was taken into a room with a photographer and several male and female assistants. After being photographed, head and shoulders, the applicant had to strip naked. "Despite their utmost reluctance," one witness recorded, "people had to undress completely . . . in order to be taken again from all sides." They were fingerprinted and measured, "during which the men obviously measured the women, hair strength was measured, blood samples taken and everything written down and enumerated."[34] Every Jew was required to go through this degradation, without exception. Some bolted as soon as they got their cards stamped, so the SS began doing the photography first.

By September the situation in Vienna was quiet, and a semblance of normal life began to resume, even for Jews within their communities.[35] But the Nazis were far from content with what they had done so far; a spur was needed to push people to the next level of Jew hatred.

In October an incident occurred in Belgium which foreshadowed what was to come. The port city of Antwerp had a large and prosperous Jewish quarter. On October 26, 1938, two journalists from the Nazi propaganda paper *Der Angriff* came ashore from a passenger steamer and began taking photographs of the Jewish diamond exchange. They behaved in an intrusive and offensive manner, and several Jews reacted angrily; they tried to eject the journalists, and there was a scuffle in which one of the Germans was hurt and their camera taken.[36] In the German press the incident was blown up into an outrageous assault on innocent and helpless German citizens. According to Vienna's main newspaper, a party of German tourists had been set upon by a gang of fifty Jewish thugs, beaten bloody, and had their property stolen as they lay unconscious. "A large part of the Belgian press is silent," the paper fumed. "This attitude is indicative of the inadequacy of these papers, which are not afraid to make a fuss when a single Jew is held accountable for his crimes."[37] The Nazi paper *Völkischer Beobachter* issued a dire warning that any further acts of Jewish violence against Germans "could easily have consequences beyond their sphere of influence, which might be extremely undesirable and unpleasant."[38]

The threat was clear, and tensions high.

As November began, anti-Semitic feelings all across the Reich were looking for an outlet. The trigger was pulled far away in Paris, when a Polish Jew called Herschel Grynszpan, in a blaze of rage over the expulsion of his people from Germany—including his own family—took a new-bought revolver into the German Embassy and fired five bullets into Ernst vom Rath, an official chosen at random.

In Vienna the newspapers called the assassination an "outrageous provocation."[39] The Jews must be taught a lesson.

Vom Rath died on Wednesday, November 9. That night, the Nazis came out in force on the streets of Berlin, Munich, Hamburg, Vienna, and every other town and city. Local party officials and the Gestapo were the masters of ceremonies, and under their lead came the SA and the SS, armed with sledgehammers, axes, and combustibles. The targets were homes and businesses still in Jewish hands. Jews were beaten and murdered out of hand if they got in the way. The storm troopers tore down and burned wherever they could, but it was the shattering of glass that onlookers remembered most vividly; the Germans called it *Kristallnacht*, night of crystal glass,[40] for the glittering shards that carpeted the pavements. The Jews would remember it as the November Pogrom.

The general order was that there was to be no looting, only destruction.[41] In the chaos that ensued the order was broken many times over, with Jewish homes and businesses robbed under cover of searching for weapons and "illegal literature."[42] Jews denounced by their neighbors had their homes invaded, possessions broken, furnishings and clothes slashed and torn by brown-shirted men; mothers shielded their terrified children and couples clung to each other in petrified despair as their homes were violated.

In Leopoldstadt, Jews caught outdoors were driven into the Karmelitermarkt and beaten. After midnight the synagogues were set ablaze, and the rooftops within sight of the Kleinmanns' apartment glowed orange, illuminated by the flames of the Polnische Schul, the synagogue in Leopoldsgasse. The fire brigade turned out, but the storm troopers barred them from fighting the fire until the magnificent building had been completely consumed. In the city center, the Stadttempel, which couldn't be burned because it adjoined other buildings, was gutted instead; its gorgeous carvings, fittings, and beautiful gold-and-white paintwork were smashed and violated, the Ark and the *bimah* thrown down and broken.

Before dawn, the arrests began. Jews in their thousands— mostly able-bodied men—were snatched from the streets or dragged from their homes by the storm troopers.

Among the first taken were Gustav and Fritz Kleinmann.

2. Traitors to the People

אבא

They were taken to the district police headquarters, an imposing building of red brick and ashlar near the Prater public park.[1] The Kleinmann family had spent many a holiday afternoon in the Prater, strolling the acres of green parkland, relaxing in the beer garden, the children delighting in the rides and sideshows of the amusement park. Now, in the gloomy winter morning, the gates were shut and the steel spiderweb of the Ferris wheel loomed over the rooftops like a threat. Gustav and Fritz passed by the park entrance without seeing it, in a truck packed with other Jewish men from Leopoldstadt.

Father and son had been reported to the storm troopers by their neighbors: by men who had been Gustav's close friends— *Du-Freunden** —men he had chatted to, smiled at, known and trusted, who knew his children and his life story. Yet without coercion or provocation they had pushed him over the cliff.

At the police station, the prisoners were unloaded and herded into a disused stable building.[2] Hundreds of men and women were in there already. Most had been taken from their homes like Gustav and Fritz, hundreds more seized the next morning while lining up outside the embassies and consulates of foreign nations, seeking escape;[3] others had been snatched randomly off the streets. A barked question: *"Jude oder*

---

* Friends close enough to call one another *du*, the intimate form of "you," rather than the formal *Sie*.

*Nichtjude?"** And if the answer was *"Jude"* or if the victim's appearance even hinted at it—into the back of the truck. Some were marched through the streets, abused and assaulted by crowds. The Nazis called this the *Volksstimme*—the voice of the people—and it howled through the streets with a sound of sirens, and in the light of dawn it went on and on; a nightmare from which there would now be no waking.

Six and a half thousand Jews—mostly men—had been taken to police stations across the city,[4] and none was fuller than the one near the Prater. The cells had overflowed with the first arrivals, and now people were crammed so tightly in the stable building they had to stand with hands raised; some were made to kneel so that newcomers could crawl over them.

Gustav and Fritz stuck together in the press. The hours wore by as they stood or knelt, hungry, thirsty, joints aching, surrounded by muttering and groans and prayers. From out in the yard came jeering and the sounds of beatings. Every few minutes, two or three people would be called from the room for interrogation. None came back.

Fritz and his father had lost track of the hours they had endured when at last the finger pointed at them and they struggled through the mass of bodies to the door. They were marched to another building and led before a panel of officials. The interrogation was held together by a glue of insults—*Jew-pig, traitor to the people, Jewish criminal*. Each prisoner was forced to identify with these calumnies, to own them and repeat them. The questions were the same for every man: *How much money have you in savings? Are you a homosexual? Are you in a relationship with an Aryan woman? Have you ever helped to perform an abortion? What associations and parties are you a member of?*

Following interrogation and review, the prisoners were assigned

* "Jew or non-Jew?"

29

to categories. Those labeled *Zurück* (return) were put back into confinement to await further processing. The ones marked as *Entlassung* (dismissal) were released—mostly women, the elderly, adolescents, and foreigners arrested by mistake. The category every man dreaded to hear was *Tauglich* (able-bodied), which meant Dachau or Buchenwald, or the new name that was being whispered: Mauthausen, a camp they were building in Austria itself.[5]

While they waited for their verdicts, Gustav and Fritz were put in a mezzanine room overlooking the yard. Here they could see the source of the noises they had heard. The men outside had been forced into packed ranks with their hands raised, lambasted and abused by storm troopers armed with sticks and whips. They were made to lie down, stand up, roll around; whipped, kicked, laughed at, their coats and good suits smeared with dirt, their hats trampled on the ground. Some were singled out for severe beatings. Those not taking part in the "gymnastics" were made to chant: "We are Jewish criminals! We are Jew-pigs!"

Throughout this, the regular police, men of long service who knew the Jewish folk of Leopoldstadt, stood by, assisting as required. Although few participated in the abuse, neither did any resist it. At least one senior policeman joined in with the beatings in the courtyard.[6]

After a long wait, Fritz's and Gustav's verdicts came through. Fritz, only fifteen years old, had been tagged *Entlassung*. He was free to go. Gustav was marked *Zurück*: back to the cells. Fritz could do nothing but watch in sick dismay as his papa was force-marched away.

בן

It was evening when Fritz left the police station. He walked home alone, passing the familiar entrance of the Prater. He'd walked this route many times before—after swimming with his friends in the Danube, after days out in the park, in a bliss of sweet cakes or buzzing with adrenaline. Now there was just emptiness.

The streets were sullen and bloodshot, hungover after the previous night's debauch. Leopoldstadt was devastated, the pavement of the shopping streets and the Karmelitermarkt carpeted with glass shards and splintered wood.

Fritz came home to the apartment, to the arms of his mother and sisters. "Where is Papa?" they asked. He told them what had happened, and that Papa had been detained. Again the terrible names pushed to the front of their minds: Dachau, Buchenwald. They waited through that night, but no word came; they enquired tentatively, but could learn nothing.

Around the world, news of the pogrom was met with revulsion. The United States recalled its ambassador from Berlin in protest,[7] the president declaring that the news "has profoundly affected the American people . . . I had difficulty believing that such things could occur in the 20th century."[8] In London *The Spectator* (then a liberal left-wing magazine) said that "barbarism in Germany is on so vast a scale, is marked by an inhumanity so diabolical and bears marks of official inspiration so unmistakable that its consequences . . . are yet beyond prediction."[9]

But the Nazis dismissed the atrocity claims as false reporting designed to distract from the real outrage—the terroristic Jewish murder of a German diplomat. They congratulated themselves on having dealt the Jews a deserved punishment, an "expression of a righteous disgust among the broadest strata of the German people."[10] Condemnations from abroad were dismissed as "dirt and filth fabricated in the known centers of immigration of Paris, London, and New York, and guided by the Jewish-influenced world press."[11] Destruction of the synagogues meant that Jews "can now no longer hatch plots against the State under cover of religious services."[12]

Fritz, Tini, Herta, Edith, and Kurt waited through that

Friday, and could discover nothing about Gustav. Then, as dusk fell and the Shabbat began, there was a knock on the door. Nervously, Tini went and opened it. And there he was—her husband, alive.

Exhausted, famished, dehydrated, gaunter than ever, Gustav walked in like a resurrection from the grave, to an outburst of joyful relief. He told his story. The Nazi officials had taken note of his service in the Great War, and old friends among the police had vouched for his many combat wounds and decorations. The standing order from the top of the SS was that veterans were excluded from the roundup, along with the sick, the elderly, and juveniles.[13] Even the Nazis wouldn't go so far yet as to condemn a war hero to a concentration camp. Gustav Kleinmann was free to go.

Over the next few days, the transports began. Fleets of *Grüne Heinrich** police vans drove in relays from police stations all over the city, packed with Jewish men—some of them war veterans too, but lacking Gustav's decorations or acquaintances in the police. They were all headed to the same destination: the loading ramp of the Westbahnhof railway station. There the prisoners were herded into freight wagons. Some went to Dachau, some to Buchenwald. Many would never be seen again.

אבא

Gustav absently twisted a strip of fabric around his fingers: an offcut, a scrap of waste, a remnant of his livelihood. The street echoed with the sound of hammering as a workman opposite drove nails into the planks covering the broken panes of a Jewish-owned shop. It was Jewish no longer.

Looking along Im Werd and across at the market and Leopoldsgasse, he picked out the businesses that had once

---

* Green Henry: equivalent to Black Maria, or "paddy wagons."

belonged to Jewish friends and were now either empty or in the hands of non-Jews. Like the neighbors who had turned him and Fritz over to the SA, many of the new owners had been friends of the people whose shops they had taken. There was Ochshorn's perfumery on the far corner of the market square, now owned by Willi Pöschl, a neighbor from Gustav's building. The butchers, poulterers, and fruit sellers had lost their market stalls: Another friend of Gustav's, Mitzi Steindl, had eagerly participated in pushing out the Jews and seizing their businesses; she'd been poor before all this, and Gustav had often given her work as a seamstress just to help her out.

With a whole class marked as enemies of the people, and the chance of an instant profit, friend had turned on friend without hesitation or qualm. Many of them reveled in the baiting, the intimidation, plundering, beatings, and deportations. In the eyes of all but a few, Jews could not be friends, for how can a dangerous, predatory animal be a friend to a human being? It was inconceivable.

An English journalist observed: "It is true that Jews in Germany have not been formally condemned to death; it has only been made impossible for them to live."[14] In the face of this impossibility, hundreds took their own lives, accepting the inevitable and relieving themselves of this hopeless nothing of a life. Many more decided to leave and find a life elsewhere. Ever since the Anschluss, Austrian Jews had been trying to emigrate, and now their numbers and their desperation increased.

משפחה

Gustav and Tini talked about leaving. Tini had relatives and friends who had gone to America many years ago. But leaving the Reich for a better place had become extremely difficult for a Jewish family without wealth or influence. In the five and a half years since the Nazis had taken power in Germany, tens of thousands of Jews had emigrated, but every nation on earth increasingly resisted the flow of migrants and refugees.

In Austria, Jewish emigration—and life generally—came under the control of Adolf Eichmann. Formerly a clerk with the intelligence and security arm of the SS, the Austrian-born Eichmann had made himself the organization's foremost expert on Jewish culture and affairs.[15] His solution to the "Jewish problem" was, first and foremost, to encourage Jews to leave, via the Central Office for Jewish Emigration. He reactivated the Israelitische Kultusgemeinde (IKG), Vienna's Jewish cultural and welfare organization, forcing its leaders to become part of his apparatus. The IKG compiled information on Jews and coordinated the bureaucracy required for their departure.

Despite wanting the Jews gone, the Nazis couldn't resist making it as cruelly hard for them as possible. They stripped them of their wealth as they passed through the system, imposing a variety of extortionate taxes and fines, including an "escaping the Reich" tax of 30 percent of their assets and an "atonement" tax of 20 percent (a punishment for Jewry's "abominable crimes"),[16] plus hefty bribes and an exchange rate for foreign currency that was pure theft. Moreover, the applicant's tax clearance was only valid for a few months, and securing a visa often took longer than that. Would-be emigrants were often flung right back to the start and had to pay all over again. As a result, the Nazi government had to *lend* the IKG money in order to help pay for impoverished Jews to obtain their travel tickets and foreign currency.[17] In this way, the Nazis' own hatred gummed up the workings of the very machine they had created to carry it out.

Finding a place to emigrate to was the hardest part. Around the world, people condemned the Nazis and criticized their own governments for doing too little to take in refugees. But the campaigners were outnumbered by those who did not want immigrants in their midst, taking their livelihoods and diluting their communities. The German press jeered at the hypocrisy of a world that made so much indignant noise about the supposedly pitiful plight of the Jews but did little or nothing to help. *The Spectator* called it "an outrage, to the Christian conscience especially, that the modern world with all its immense wealth and resources cannot give these exiles a home."[18]

For the Kleinmann family, their city had become, in the words of a British journalist:

> . . . a city of persecution, a city of sadism . . . no amount of examples of cruelty and bestiality, can convey to the reader who hasn't felt it the atmosphere of Vienna, the air which the Austrian Jews must breathe . . . the terror at every ring of the front-door bell, the smell of cruelty in the air . . . Feel that atmosphere and you can understand why it is that families and friends split up to emigrate to the corners of the earth.[19]

Even after Kristallnacht, foreign governments, the conservative press, and the prevailing democratic will continued to stand firm against letting in more than a trickle of Jewish migrants. When people in the West looked to Europe, they saw not only the few hundred thousand Jews in Germany and Austria, but looming behind them the thousands in other Eastern European countries, and the three million in Poland; all these nations had recently enacted anti-Semitic laws.

"It is a shameful spectacle," said Adolf Hitler, "to see how the whole democratic world is oozing sympathy for the poor tormented Jewish people but remains hard-hearted and obdurate when it comes to helping them."[20] Hitler sneered at Roosevelt's "so-called conscience," while in Westminster MPs from all parties spoke earnestly about the need to help the Jews, but Home Secretary Sir Samuel Hoare warned of "an underlying current of suspicion and anxiety about an alien influx" and advised against mass immigration.[21] However, the members, prompted by Labour MPs George Woods and David Grenfell, insisted on a concerted move to help Jewish children—to save "the young generation of a great people" who "have never failed . . . to make a handsome and generous contribution" to the way of life of nations that gave them asylum.[22]

Meanwhile, Jews in the Reich could only live out their days, queue at the consulates of Western nations, and wait and hope that their applications would be successful. For the thousands in concentration camps, an emigration visa was their only hope. Hundreds in Vienna were homeless, and many were reluctant to apply to emigrate for fear of arrest.[23]

Gustav had no money and no property, so he couldn't raise the funds to buy his way through the bloodsucking bureaucracy. He also felt little confidence in his ability to begin a new life in a strange country. The final word lay with Tini, who simply couldn't bear the thought of leaving. She was rooted in Vienna, born and bred. At her age, where could she possibly go without feeling torn from her natural place? Her children were another matter. She was especially worried about fifteen-year-old Fritz; the Nazis had taken him once, and might take him again. It wouldn't be long before he lost the protection of his age.

In December 1938, over a thousand Jewish children left Vienna for Britain—the first of a projected five thousand accepted by the UK government, living up to its fine words for once.[24] Eventually over ten thousand would find safety in Britain through the *Kindertransport*. Yet even this was just a fraction of those needing refuge. The British proposed opening up Palestine to ten thousand additional children. Tini heard about this proposal and had hopes of placing Fritz on one of the transports;[25] he was old enough to cope with being sent away and to support himself through work, which eight-year-old Kurt could not. The talks in Palestine dragged on for months. The Arabs feared being swamped in their own land, losing the majority rights they currently enjoyed and sacrificing all their hopes of a future independent Palestinian state. The talks eventually broke down.[26]

While the rest of her family fretted and wavered, Edith Kleinmann was absolutely determined to leave. On top of the degradation and abuse she had suffered, as a lively, outgoing spirit she couldn't bear this confinement, which amounted to a kind of captivity. Whatever it took she had to get out.

Edith had her eyes on America, and had acquired the two affidavits she needed from her mother's relatives there, who were willing to provide her with shelter and support. Thus prepared, at the end of August 1938 she had registered at the American consulate to begin the application process.[27] The system was bursting with applicants, and was deliberately squeezed tight at both ends, by the State Department and the Nazi regime. With the end of the year looming, Edith faced the prospect of being stuck in Vienna forever. After Kristallnacht, impatient with waiting, she decided that England looked like a better prospect.

Since the early summer, large numbers of Jews—mostly women, who passed more easily through the vetting process—had fixed on Britain as the place to try for. Hopeful advertisements had begun to appear in the classified section of *The Times*.[28] The advertisers ranged from maids, cooks, chauffeurs, and nannies to goldsmiths, doctors of law, piano teachers, mechanics, language tutors, gardeners, and bookkeepers. Many offered themselves for more lowly work than they were qualified for. The same self-recommendations recurred: "good teacher," "perfect cook," "good handyman," "experienced," "excellent character." As time went on, the advertisements became palpably desperate: "any work," "urgently seeks," "with boy aged 10 (in children's home if necessary)," "immediately" . . . the clamoring of people with prison walls rising around them and doors slamming shut.

Certified domestic servants had the best chance of getting a visa.[29] A near neighbor of the Kleinmanns, Elka Jungmann, placed an ad that was typical of the hundreds of others:

COOK, with long-service testimonials (Jewess), also housekeeper, knows all housework, seeks post.—Elka Jungmann, Vienna 2, Im Werd 11/19.[30]

As an apprentice milliner, Edith had no domestic skills to offer, and she wasn't keen on acquiring any. She dressed well, lived well, and saw herself as a lady. Clean the house? It wasn't in her nature. But Tini took her in hand, teaching her what she could, and obtained her a placement as a maid with a well-off local Jewish family. Edith worked there for one month, and they generously gave her a testimonial certifying that she had worked for six. With amazing good fortune Edith managed to obtain a work contract in England. All she needed now was a visa and clearance from the Nazi authorities.

This was the hard part. The British government gave out only a handful of visas each day.[31] The line at the British consulate was long and painfully slow. Twenty-four hours a day, all the members of the family took turns holding Edith's place in line. The cold was bitter, but they kept to their turns as the line inched forward day by day. The streets outside the various consulates were clogged with applicants, who were periodically dispersed by the police; sometimes SA men would come by and beat the Jews with rope ends.[32] It took a whole week for Edith's place to reach the grand doorway of the Palais Caprara-Geymüller, which housed the British consulate.[33] She was admitted, and lodged her application. Then she waited. At last, in early January 1939, she was granted her visa.

Edith's parting was painful for everyone. None of them could imagine how or when they would ever meet again. She boarded a train and vanished from their lives into a new existence, leaving a void in the family.

Within days Edith was aboard a ferry crossing the English Channel, leaving the terror and the abuse and the danger behind her, but also everything she knew and everyone she loved, fearful for what might happen to them. In later years, when she grew old and talked to her children about this time, she would fall silent at this point, as if the pain still remained too sharp, long after all else had lost its bite—this memory of parting more potent than anything that had gone before.

משפחה

In Vienna, the besieged Jewish community was a ghost of its former self. A visitor who came in the early summer of 1939 believed it was worse than anything in Germany; whole streets of shops and houses in Leopoldstadt were left vacant where Jews had been evicted; formerly busy streets were deserted, "and it looked to us just like a dead city."[34]

The Zionist Youth Aliyah, whose official purpose was to prepare young Jews for kibbutz life in Palestine, did heroic work among the children, providing teaching, training in crafts and medicine, and succor. Over two-thirds of Vienna's remaining Jews now depended on charity, mostly within their own communities. They went outdoors as little as they could. In most districts it was dangerous for them to be out after dark, especially on evenings when Nazi Party meetings took place; there would always be some brutality after the SS and SA had wound themselves up with speeches. Some districts were too dangerous at any time of day or night.

In their apartment, the Kleinmann family held together, closing in around the empty space left by Edith. Kurt attended one of the improvised schools, while his brother and sister did what they could to help their parents. That summer, Fritz turned sixteen and had to get a new identity card. Of all the family's *J-Karte* photos, Fritz's—in which the good-looking boy, dressed only in his undershirt, glared with detestation into the camera—was the only one that would ultimately survive.

Occasional letters found their way to Vienna from Edith. They were short and simple. Edith had settled into her work as a maid, and was doing well. She lived in the suburbs of Leeds and worked for a Russian Jewish lady called Mrs. Brostoff. She said nothing of her feelings.

Edith's letters continued to arrive during that summer, then abruptly stopped; on September 1, Germany invaded Poland. Britain and France declared war, and an impenetrable barrier fell between Edith and her family.

Nine days later, an even worse blow fell upon them. On September 10, Fritz was seized by the Gestapo.

אבא

A new wave of arrests was sweeping through the Reich. With Germany at war with Poland, all Jews of Polish origin were classed as enemy aliens.[35] As an Austrian citizen, born and bred, Gustav should have been safe. However, people who knew him well were aware that he'd been born in the old kingdom of Galicia. Since 1918, Galicia had become part of Poland, and as far as Germany was concerned, any Jew born there was Polish and a security threat.

The hammer blow fell on a Sunday, when Tini was in the apartment with Herta, Fritz, and Kurt. There was a loud knock at the door, making them all flinch in terror.

Tini opened the door warily and peeped out. Four men loomed over her, all neighbors. She recognized every face; every line under the eyes and every bristle on their cheeks was a familiar sight. All were working men like Gustav—friends with wives she knew, whose children had once played with hers. There was Friedrich Novacek, an engineering worker, and foremost among them Ludwig Helmhacker, a coalman.[36] These were the same men who had turned Gustav over to the authorities on Kristallnacht, and Ludwig and his little gang of Nazi quislings had called many times since.

"What do you want from us now, Wickerl?" Tini said in exasperation as they pushed past her into the small apartment. (Despite everything, she couldn't help calling Ludwig by the familiar diminutive.) "You know we've got nothing—we don't even have food."[37]

"We want your husband," said Ludwig. "We have orders; if Gustl* isn't here, we're supposed to take the lad." He nodded at Fritz.

Tini felt as if she'd been physically kicked. There was nothing she could say to change what was happening. They took hold of her precious boy and marched him out of the door. Ludwig paused before leaving. "See, we'll take Fritzl to the police, and when Gustl reports, the lad can come home again."

When Gustav returned later that day, he found his family in a state of panic and grief. When he heard what had happened, he didn't hesitate; he turned right round and headed for the door, intending to go straight to the police. Tini grabbed his arm. "Don't," she said. "They'll take you."

"I'm not leaving Fritzl in their hands." He made for the door again.

"No!" Tini pleaded. "You have to run away, go somewhere and hide."

There was no shaking him. Leaving Tini in tears, Gustav walked quickly to the police station in Leopoldsgasse. Taking

---

* Affectionate diminutive used in eastern Austria; e.g., Fritzl, Gustl.

his courage in both hands, he walked right in and up to the desk. The police officer on duty looked up at him. "I'm Gustav Kleinmann," he said. "I'm here to turn myself in. You have my son. Take me and let him go."

The policeman glanced around. "Get out," he muttered. "Get the hell out of here."

Bewildered, Gustav left the building. He went home to find Tini both relieved to see him and distraught that Fritz was still gone. "I'll try again tomorrow," he said.

"They'll come for you before then," she said. Again she pleaded with him to run away and hide, but he refused. "Get out now," she insisted, "or I'll turn on the gas—I'll kill myself." Kurt and Herta watched in horror. Their parents' resilience was the mainspring of the family; to see them reduced to despair was appalling.

Eventually Tini got through to Gustav. He left the apartment, promising to find a place to hide. All that day and evening Tini waited on tenterhooks, listening for the knock on the door. It didn't come; instead, late that night, Gustav himself returned. He had nowhere else to go, and couldn't bear to leave Tini and the children alone all night. There was no knowing who might be taken next if the Nazis didn't find him.

At two o'clock in the morning they came—the thundering on the door, the tide of men surging into the apartment, the snapped orders, the hands seizing hold of Gustav, the weeping, pleas, the last desperate words between husband and wife. He was allowed to pack a small bundle of clothes—a sweater, a scarf, a spare pair of socks.[38] And then it was all over. The door slammed, and Gustav was gone.

PART II
Buchenwald

3. Blood and Stone: Konzentrationslager Buchenwald

אבא

Making sure he was alone, Gustav took out a little pocket notebook and pencil. He wrote in his clear, angular hand: "Arrived in Buchenwald on the 2nd October 1939 after a two-day train journey."

Over a week had passed since that dreadful arrest, and a lot had happened. Even the most concise account would eat up the notebook's precious leaves. He'd managed to keep it concealed, knowing it would be the death of him if he were found with it. There was no telling whether he would ever get out of this place. Whatever happened, this diary would be his witness.

He smoothed down the page and continued writing "From Weimar railway station we ran to the camp . . ."

בן

The wagon door groaned open, flooding the inside with light; instantly a hell's chorus of shrieked orders and snarling guard dogs erupted. Fritz blinked and looked around, stunned by the barrage on his senses.[1]

It seemed an age since Wickerl Helmhacker and his pals had torn Fritz away from his mother. The only thing he had to console him was that, since he hadn't been released, that must mean his papa had got away safely.

Fritz had been taken initially to the Hotel Metropole, headquarters of the Vienna Gestapo. Huge numbers of Jewish men had been arrested and the SS were struggling to accommodate them. After a few days in the Gestapo cells, Fritz was transferred with thousands of others to the football stadium near the Prater. There they were kept under guard in crowded, unsanitary conditions for nearly three weeks. Eventually they were taken to the Westbahnhof and loaded into cattle wagons.

The journey to Germany dragged on for two days. Fritz, confined in the press of bodies, was rocked by the jolting train and oppressed by the proximity of strangers, a sixteen-year-old boy among a crowd of anxious, sweating men. They were of every kind imaginable: the middle-class father, the businessman, the bespectacled intellectual, the bristle-cheeked workman, the ugly, the handsome, the portly, the terrified, the man who took it all calmly, the man simmering with indignation, the man scared to his bowels. Some were silent, some muttered or prayed, some chattered incessantly. Each man an individual with a mother, a wife, children, cousins, a profession, a place in the life of Vienna. But to the men in uniforms outside the wagon—just livestock.

"Out, Jew-pigs—now! Out-out-out!"

Out they came into the dazzling light. One thousand and thirty-five Jews—bewildered, seething, confused, scared, dazed—pouring down from the cattle wagons on to the loading ramp of Weimar station into a hailstorm of abuse and blows and snarling dogs.[2] A crowd of local people had turned out to watch the transport come in; they stood beyond the SS guards, jeering, smirking, calling out insults.

The prisoners—many carrying bags and bundles and even suitcases—were pushed, beaten, and yelled into ranks. From the loading ramp they were herded into a tunnel, then out into the air again, driven along at a run. The crowd followed for a while along the northbound city street.

"Run, Jew-pigs, run!"

Fritz forced his cramped limbs to run. If any man faltered, turned aside, even looked like he was slackening his pace, or if he spoke to another, the hammer blow of a rifle butt would fall on his shoulders, his back, his head.

These SS men were worse than any Fritz had seen in Vienna; they belonged to the *Totenkopfverbände*—Death's Head units; their caps and collars bore skull-and-crossbones badges and their brutality was beyond all human reason. Drunkards and sadists with stunted or twisted minds, deformed souls—vested with a sense of destiny and almost limitless power, trained to believe that they were soldiers in a war against the enemy within.

Fritz ran and ran into a seemingly endless hell of violence. The city street gave way to mile after mile of country road. The prisoners were mocked and spat on. Men who stumbled, weakened by age or fatigue or the burden of their luggage, were immediately shot. A man might stoop to tie a shoelace, or fall over, plead for water, and he would be gunned down without hesitation. The road, climbing a long slope, led into a thick forest. There the prisoners were turned aside onto a new concrete road. Veterans called it the Blood Road. Many prisoners had died making it, and their blood was joined by that of the new arrivals driven along it.

As he ran, lungs bursting, Fritz thought he recognized a familiar tall, lean figure ahead of him. Increasing his pace, Fritz drew level. He had been right—here, in spite of all reason, was his papa! Laboring along, dripping with sweat, with Tini's little package of spare clothing under his arm.

To Gustav, it was as if Fritz had materialized out of nowhere. This was no occasion for astonishment or emotional reunions. Keeping their mouths shut and sticking close together, they edged deeper into the pack to avoid the random blows, shutting their minds to the sporadic gunshots, and ran on with the herd, up the hill, deeper and deeper into the forest.

The hill was the Ettersberg, broad-backed and covered in dense beech woods. For centuries it had been a hunting ground of the dukes of Saxony-Weimar, and more recently a popular spot for picnics. It had been a retreat for artists and intellectuals, famously associated with writers like Schiller and Goethe.[3] The city of Weimar was the very epicenter of German classical cultural heritage; by founding a concentration camp on the Ettersberg, the Nazi regime was placing its own imprint upon that heritage.

At last, after five miles, which had taken the prisoners more than an hour to run, the Blood Road bent northward and emerged into a vast open space cleared in the forest. Scattered across it were buildings of all shapes and sizes, some complete, some still under construction, many hardly begun. These were the barracks and facilities of the SS, the infrastructure of the machine in which the prisoners were both fuel and grist. Buchenwald—named for the picturesque beech forest which made the mountain so pleasant—was more than just a concentration camp; it was a model SS settlement whose scale would eventually rival that of the city itself. What happened here among the beeches would one day cast all of Weimar's Germanic heritage in shadow. Many of the people imprisoned here called it not Buchenwald but Totenwald—Forest of the Dead.[4]

Ahead the road was barred by a wide, low gatehouse in a massive fence. This was the entrance to the prison camp itself. On the gateway were two slogans. Above, on the lintel, was inscribed:

RECHT ODER UNRECHT—MEIN VATERLAND

My country, right or wrong: the very essence of nationalism and fascism. And wrought into the ironwork of the gate itself:

JEDEM DAS SEINE

To each his own. It could also be read as *Each person gets what he deserves.*

Exhausted, sweating, bleeding, the new arrivals were herded through the gate. There were now 1,010 of them; twenty-five of the men who had set out from Vienna were now corpses along the Blood Road.[5]

They found themselves within an impenetrable cordon: the huge camp was surrounded by a barbed-wire fence with twenty-two watchtowers at intervals, decked with floodlights and machine guns; the fence was ten feet high and electrified, with a lethal 380 volts running through it. The outside was patrolled by sentries, and inside was a sandy strip called the "neutral zone"; any prisoner stepping on it would be shot.[6]

Immediately inside the gate was a large parade ground—the *Appelplatz*, or roll-call square. Ahead and along one side were barrack huts that marched in orderly, radiating rows down the hill slope, with bigger two-story blocks beyond. Gustav and Fritz and the rest of the newcomers were ordered into ranks in the roll-call square. They stood at gunpoint, awkward and disheveled in their soiled business suits and work clothes, sweaters and shirts, raincoats, fedoras and office shoes, caps and hobnail boots, bearded, bald, slicked hair, tousled mops. While they stood, the bodies of the men murdered along the road were carried in and dumped among them.

A group of finely dressed SS officers appeared. One, a middle-aged, pouchy-faced man with a slouching posture, stood out. This, they would learn later, was Camp Commandant Karl Otto Koch. "So," he said, "you Jew-pigs are here now. You cannot get out of this camp once you are in it. Remember that—you will not get out alive."

The men were entered in the camp register, and each assigned a prisoner number: Fritz Kleinmann—7290; Gustav Kleinmann—7291.[7] Orders came in a confusing barrage which many of the Viennese found hard to understand, unaccustomed to German dialects. They were made to strip naked and

march to the bath block, where they showered in almost un-
bearably hot water (some were too weak to stand it, and collapsed). Then came immersion in a vat of searing disinfectant.[8]
Naked, they sat in a yard to have their heads sheared, and under yet another rain of blows from rifle butts and cudgels were
made to run back to the roll-call square.

There they were issued with camp uniforms: long drawers,
socks, shoes, shirt, and the distinctive blue-striped trousers
and jacket, all ill-fitting. If desired, for twelve marks a prisoner
could buy a sweater and gloves,[9] but few had so much as a pfen-
nig. All their own clothes and belongings—including Gustav's
little package—were taken away.

Scalps shaved, dressed in uniform, the new arrivals were no
longer individuals but a homogeneous mass identified only by
their numbers, the only distinguishing features an occasional
fat belly or a head standing higher than the rest. The violence
of their arrival had impressed on them that they were the property of the SS, to do with as it saw fit. Each man had been issued with a strip of cloth with his prisoner number on it, which
he was required to sew onto the breast of his uniform, along
with a symbol. Examining his, Fritz saw that it was a Star of
David made up of a yellow and a red triangle superimposed.
All the other men had the same. The red triangle denoted that,
having been arrested on the pretext that they were Jewish-
Polish enemy aliens, they were under so-called "protective custody" (meaning "protection" for the state).[10]

The prisoners were now inspected by another SS officer,
who had a flat face like the back of a shovel. This, they would
learn, was Deputy Commandant Hans Hüttig, a dedicated sadist. Surveying them with disgust, he shook his head and said,
"It's unbelievable that such people have been allowed to walk
around free until now."[11]

They were marched to the "little camp," a quarantine enclosure on the western edge of the roll-call square surrounded by a double cordon of barbed wire. Inside, rather than barrack huts, were four huge tents containing wooden bunks four tiers high.[12] In recent weeks, over eight thousand new prisoners had arrived at Buchenwald, more than twenty times the usual rate of intake,[13] and the tents were full to bursting.

Gustav and Fritz found themselves sharing a bunk space only six feet wide with three other men. There were no mattresses, just bare wooden planks. They had a blanket each, so they were at least warm. Squeezed in like sardines and their bellies empty, they were so dead tired they fell asleep right away.

The following day, the new prisoners were registered with the camp Gestapo—photographed, fingerprinted, and briefly interrogated, a process that took all morning. In the afternoon they received their first warm food: a pint of soupy stew containing unpeeled potatoes and turnips, with a little fat and meat floating in it. The evening meal consisted of a quarter-loaf of bread and a little piece of sausage. The bread was provided in whole loaves, and as there were no knives, sharing it out was a haphazard business which usually led to disputes and jealous quarrels.

For eight days they were left in quarantine, then put to work. Most were set to hard labor in the nearby stone quarry, but Gustav and Fritz were employed on maintaining the canteen drains. All day long the workers were abused and slave driven. Gustav wrote in his diary: "I have seen how prisoners get beaten by the SS, so I look out for my boy. It's done by eye contact; I understand the situation and I know how to conduct myself. Fritzl gets it too."

So ended his first entry. He looked back over what he had written so far, just two and a half pages to bring them this far, through this much distress and danger. Eight days gone. How many more to come?[14]

אבא

Gustav understood that to stay safe it was vital to remain unnoticed. But within two months of arriving in Buchenwald, both he and Fritz drew attention to themselves in the most dangerous way possible—Gustav unwillingly, Fritz deliberately.[15]

Each morning, an hour and a half before dawn, shrill whistles yanked them from the forgetfulness of sleep. Then came the kapos and the block senior, yelling at them to hurry. These men were a shock to new arrivals; they were fellow prisoners—mostly "green men," criminals who wore the green triangle on their uniforms—appointed by the SS to act as slave drivers and barrack overseers, allowing the SS guards to keep a distance from the mass of prisoners.

As the whistles shrilled, Fritz and Gustav put on their shoes and scrambled down, sinking to their ankles in cold mud on the bare floor. Outside, the camp was ablaze with electric light along the fence lines, atop the guard towers, and in the walkways and open areas. They were herded to the square for roll call, receiving a cup of acorn coffee each. It was sweet but had no power to stimulate, and was always cold by the time they got it. Doling it out was a long process, and they had to stand in silence, motionless and shivering in their thin clothes for two hours. When it was time to go to work, sunrise was beginning to lighten the landscape.

Gustav and Fritz had only enjoyed a brief stint on drainage work, and had now been assigned to the quarry detail. Forming orderly columns, they were marched out through the main gate, turning right to follow the road leading down between the main camp and the SS barrack complex—a set of large two-story brick buildings, some still under construction, arranged in an arc like the blades of a fan. The Nazis adored their grand designs, even in their concentration camps—an illusive appearance of elegance, order, and meaning to screen the nightmare.

A little way down the hill, the prisoners passed through the inner sentry line. There were no fences outside the main camp, and the work areas were surrounded by a well-manned cordon of SS sentries. They were spaced at forty-foot intervals; every second sentry was armed with a rifle or sub machine gun, and every other with a cudgel. Once inside the sentry line, any prisoner who crossed it was shot without hesitation or challenge. For the desperate, running into the line was a common means of suicide. For certain SS guards, forcing prisoners to run over the line was a favorite means of entertainment. An "escape register" was maintained, recording the names of the SS marksmen and awarding credit for kills, which added up to rewards of vacation time.

The quarry was large—a pale, raw limestone scar on the green wooded hillside. From it, if one raised one's head, and the mist and rain permitted, a vista of broad, rolling countryside stretched to the hazy western horizon. But one didn't raise one's head—not for more than a moment. The work was hard, ceaseless, dangerous; the men in stripes dug stone, broke stone, carried stone, and were beaten by the kapos if they slacked. A kapo was expected to be harsh, motivated by the knowledge that if the SS were dissatisfied with him, they would remove his status and put him back among the prisoners, who would exact their revenge.[16]

There was a narrow railway in and out, on which huge steel dump wagons ran, each the size of a farm cart, carrying the stone from the quarry to the construction sites around Buchenwald. Gustav and Fritz worked as wagon haulers; all day, they and fourteen other men had to heave and push a laden wagon weighing around four and a half tons[17] up the hill, a distance of half a mile, lashed and yelled at by kapos. The rails were laid on beds of crushed stone, which slipped and grated under the men's flimsy shoes or wooden clogs. Speed was imperative, and as soon as the wagon was emptied, it had to get back to the quarry at once, running down the return track, propelled by its own weight, with the sixteen men holding on to prevent it racing out of control. Falls were frequent, with fractured limbs and broken heads. Often a wagon would jump the rails, sometimes directly in the path of the next wagon, leaving a trail of men crushed and dismembered.

The injured would be taken off to the infirmary; or, if they were Jews, to the Death Block—a holding barrack for the terminally sick.[18] Men with crippling injuries would be given a lethal injection by an SS doctor.[19] Even slight wounds could be life-threatening in the unsanitary conditions in which the prisoners lived and worked. For a man with poor eyesight, losing his glasses could effectively be a death sentence.

Gustav and Fritz toiled on day after day, managing to avoid both punishment and injury. "We are proving ourselves," Gustav wrote in his diary.

So it went on for two weeks. Then, on October 25, dysentery and fever broke out in the quarantine camp. They had no water supply, and workers in the quarry would drink from puddles; this was believed by some to be the cause. With over 3,500 weakened men crammed into its tents, and sanitation consisting of nothing but a latrine pit, it was a fertile ground for disease. Each day the population was eroded by dozens of deaths.

Nevertheless, the grinding life of the camp went on. Each day, impoverished rations; each day, standing for hours at roll call in the cold and rain; each day, beatings and injuries. The SS waged a special vendetta against a chief rabbi called Merkl, who was regularly beaten bloody, and eventually forced to run through the sentry line. And all the while the dysentery went unchecked and the death toll rose.

Some Poles, driven by hunger, cut their way out of the little camp and broke into the main camp kitchens, returning with three gallons of syrup, a delight that slightly brightened the prisoners' diets. It was a short-lived pleasure. The theft was discovered, and the whole of the little camp was punished with two days' withdrawal of rations. A few days later, a crate of jellied meat was stolen from the store. The prisoners were starved again for two days and forced to stand at attention on the roll-call square from morning until evening. While the punishment parade was still going on, there was a break-in at the piggery in the farm site at the north end of the camp, and a pig was taken. Camp Commandant Koch (who lived in a pleasant house in the Buchenwald complex and went for Sunday walks in Buchenwald's own zoo with his wife and little children) personally ordered starvation for everyone until the thieves were caught. Every prisoner's clothing was inspected for signs of blood or sawdust from the pigpen. The punishments and interrogations went on for three days, until it was finally discovered that the thieves were actually SS men.[20]

Weakened by starvation, subjected to soul-breaking labor, the living walked silent and hunched like specters of the already dead.

Then, suddenly, things got even worse.

בן

On Wednesday, November 8, 1939, Adolf Hitler flew to Munich to lead the Nazi Party's annual commemoration of the failed 1923 Beer Hall Putsch, when he and his followers made their first attempt to seize power in Bavaria. Hitler opened the occasion with a speech in the grandiose Bürgerbräukeller beer hall. With the war only just begun and his planned invasion of France facing postponement due to bad weather, the Führer planned to rush back to Berlin, and therefore gave his address an hour earlier than scheduled. Eighteen minutes after his departure—when he should have been in the middle of his speech—a bomb concealed in a pillar exploded with colossal force, obliterating the handful of people standing nearby and injuring dozens of others.[21]

Germany was appalled. Although the perpetrator, Georg Elser, was a German communist with no Jewish connections, in Nazi eyes the Jews were responsible for every ill deed. In the concentration camps next day—which happened to be the anniversary of Kristallnacht—they took brutal revenge. In Sachsenhausen, the SS subjected the inmates to intimidation and torture, while at Ravensbrück, the Jewish women were locked in their barracks for nearly a month.[22] But these cruelties paled beside what occurred at Buchenwald.

Early in the morning of November 9, all Jewish prisoners, including Gustav and Fritz, were taken from their work details and marched back to the main camp. They were ordered back to their barrack blocks, and when all were confirmed present and correct, SS-Sergeant Johann Blank commenced the ritual punishment.

Blank was a born sadist. A former forestry apprentice and poacher from Bavaria, he was an enthusiastic participant in the game of forcing prisoners across the sentry line, carrying out many of the murders personally.[23] Accompanied by other SS men, still hungover from the previous night's Putsch celebrations, Blank went from block to block picking out twenty-one Jews, including a seventeen-year-old boy who had the bad luck to be outdoors on an errand. They were marched to the main gate, where they had to stand while the SS men performed a little parade to coincide with the commemorative march then taking place in Munich. When they were finished, the gate was opened and the twenty-one Jews were herded down the hill toward the quarry.

Inside their tent, Gustav and Fritz knew nothing of what was going on, other than the sounds that carried their way. For a long while, there was silence. Then, suddenly, there came a crackle of gunfire; then another and another, followed by sporadic shots. Then silence again.[24]

Reports of what had happened quickly circulated round the camp. The twenty-one had been marched to the quarry entrance, where they had all been shot. A few had managed to run, only to be hunted down and murdered among the trees.

The day wasn't over yet. SS-Sergeant Blank, accompanied by Sergeant Eduard Hinkelmann, now turned their attention to the little camp. They carried out an inspection of the tents, finding fault with everything and working themselves into a fury. They ordered the prisoners out to the roll-call square. When they were lined up, the kapos went along the ranks, grabbing every twentieth man and shoving him forward. They came along Gustav and Fritz's line: *one, two, three* . . . the counting finger danced along, pulsing the beats . . . *seventeen, eighteen, nineteen*: the finger went past Gustav . . . *twenty*: the finger jabbed at Fritz.

He was seized and pushed toward the other victims.[25]

A heavy wooden table with straps dangling from it was dragged onto the square. Any prisoner who had been here more than a week or two recognized the *Bock*—the whipping bench. It had been introduced by Deputy Commandant Hüttig as a means of punishing prisoners and entertaining his men.[26] Every prisoner had witnessed its use and was terrified by the sight of it. Sergeants Blank and Hinkelmann very much enjoyed putting it to work.

Fritz was gripped by the arms and, with his insides

dissolving, was rushed to the *Bock*. His jacket and shirt were removed and his trousers pulled down. Hands shoved him facedown on the sloping top, forced his ankles through the loops, and tightened the leather strap over his back.

Gustav watched in helpless dismay as Blank and Hinkelmann prepared; they relished the moment, stroking their bullwhips—ferocious weapons of leather with a steel core. Camp regulations allowed for a minimum of five lashes and a maximum of twenty-five. Today the rage of the SS could be sated by nothing less than the full number.

The first lash landed like a razor cut across Fritz's buttocks.

"Count!" they yelled at him. Fritz had witnessed the ritual before; he knew what was expected. "One," he said. The bullwhip cut across his flesh again. "Two," he gasped.

The SS men were methodical; the lashes were paced to prolong the punishment and heighten the pain and terror of each blow. Fritz struggled to concentrate, knowing that if he lost count the lashes would start over again. Three . . . four . . . an eternity, an inferno of pain . . . ten . . . eleven . . . fighting to concentrate, to count correctly, not to give in to despair or unconsciousness.

At last the count reached twenty-five; the strap was loosed and he was forced to his feet. Before his father's eyes he was helped away, bleeding, on fire with pain, his mind stunned, as the next unfortunate was dragged to the *Bock*.

The obscene ritual dragged on for hours: dozens of men, hundreds of slowly paced blows. Some men succumbed to the distress of the moment, miscounted their strokes and had to begin again. None walked away unbroken.

אבא

There was no medical treatment for Jews, no time off, no healing interlude. The victims, cut up and in horrible pain, were thrown immediately back into the daily routine of the camp. They had to soldier on as best they could, because to succumb to pain or sickness here was to give in to death. In Buchenwald, no matter how bad things got, they could always get worse, and regularly did.

Two days passed, and at morning roll call Fritz stood to attention with considerable difficulty. Despite his pain, he was more worried about his papa than about himself; Gustav wasn't well at all. The starvation punishment had been renewed; dysentery and fever still plagued the camp, and now Gustav had caught the sickness. He was pale, feverish, and afflicted by diarrhea. Fritz watched him from the corner of his eye as the minutes ticked slowly by. In this state he couldn't possibly work; he could scarcely stand through roll call.

Gustav swayed, shivering, his senses withdrawing. Sounds grew faint and muffled, a black haze closed in around his vision, his extremities became suddenly numb, and he felt himself falling, falling, into a black pit. He was unconscious before he hit the ground.

When he woke, he was on his back. Somewhere indoors. Not the tent. Above him floated the face of Fritz. Another man too. Was this the infirmary? That was impossible; the infirmary was closed to Jews. In his hazy, febrile state, Gustav realized dimly that this must be the block set aside for hopeless cases, from which people rarely emerged alive. The Death Block.

Fritz and the other man had carried him here—Fritz struggling despite his injuries. The air was thick, stifling, filled with groans and an atmosphere of hopeless, helpless death.

There were two doctors. One, a German named Haas, was callous and stole from the sick, leaving them to starve. The other was a prisoner, Dr. Paul Heller, a young Jewish physician from Prague. Heller did the best he could for his patients with the meager resources the SS provided.[27] Gustav lay helpless for days, with a temperature of 101°F, sometimes lucid, sometimes in a fever dream.

Fritz, meanwhile, was growing ever more worried about conditions in the little camp. They were being starved again. The announcement on the loudspeaker had been heard so many times it was like a mantra—"Food deprivation will be imposed as a disciplinary measure." This month alone they had endured eleven days of it. Some of the younger prisoners suggested begging the SS for food. Fritz, who had scarcely begun recovering from his whipping, was among them. The older, wiser prisoners, many of them veterans of the First World War, warned against it. Taking action meant exposure, and exposure meant punishment or death.

Fritz talked it over with a Viennese friend, Jakob Ihr—nicknamed "Itschkerl"—a boy from the Prater. "I don't care if we have to die," said Itschkerl. "I'm going to speak to Dr. Blies when he comes."

Ludwig Blies was the camp doctor; although hardly a kind man, he was more humane—or at least less callous—than some other SS doctors. He had on rare occasions intervened to halt excessive punishments.[28] Blies also seemed an approachable figure: middle-aged and disarmingly comical in appearance.[29]

"All right," said Fritz. "But I'm coming with you. And I'll do the talking; you just back me up."

When Dr. Blies made his next inspection, Fritz and Itschkerl diffidently presented themselves to him. Fritz, being careful not to seem demanding, made his voice quaver with weepy desperation. "We have no strength to work," he pleaded. "Please give us something to eat."[30]

Fritz had weighed his words carefully; rather than seeking pity, he was appealing to the practical SS view of prisoners as a labor resource. But it was also extremely dangerous to appear unable to work; uselessness meant death.

Blies stared in astonishment. Fritz was small for his age, and appeared little more than a child. With the effects of injury and starvation, he made a pitiful sight. Blies wavered, his humanity vying with his Nazi principles. Abruptly he said, "Come with me."

Fritz and Itschkerl followed the doctor across the square to the camp kitchens. Commanding them to wait, Blies went into the food store, and came out a few minutes later carrying a large loaf of ration-issue rye bread and a half gallon of soup. "Now," he said, handing over this astonishing bounty, "back to your camp. Go!"

They shared the food—equivalent to a meal ration for half a dozen men—with their closest bunkmates. The following day the whole camp was put back on full rations, apparently on Blies's orders. The two boys were the talk of the camp, and from that day forward Itschkerl became one of Fritz's best friends.

As the days wore on, Fritz visited his papa in the Death Block whenever he could. The dysentery had failed to kill him, and the worst was past; however, it was obvious to Gustav that he would never get well in this unsanitary, pestilential environment. After two weeks, he begged to be discharged, but Dr. Heller wouldn't let him go. He was far too weak to survive.

Gustav was determined; disobeying the doctor's orders, he asked Fritz to help him to his feet. Father and son slipped out of the Death Block together. The moment he was out in the fresh air Gustav began to feel better; with his arm around Fritz's shoulders, they made their way back to the little camp, Fritz guiding his papa's faltering steps.

Even in the muddy, overcrowded tent the atmosphere felt fresher than in the ward, and Gustav began to regain his strength. The following day he was given light work as a latrine cleaner and furnace stoker;[31] he ate better, and recovered his health a little.

Fritz too was recovering from his injuries. But there was always a limit to one's health in Buchenwald. They were both thin; Gustav, who had always been lean, had dwindled to ninety-nine pounds during his illness. Fritz's new reputation for cleverness had made him popular not only with the ordinary prisoners but even with some of the camp seniors—the very highest of the prisoner functionaries. But still the reality remained: any perks were minimal and the consolations little more than a stay of death. "I work to forget where I find myself," Gustav wrote.

With their first camp winter beginning to set in, he and Fritz were grateful to receive a parcel of fresh underwear from home. They were allowed to receive such things, but could send out no communications. A letter accompanied the parcel. Tini was trying to arrange for the children—including Fritz—to go to America, but making little progress against the bureaucratic tide. Of Edith there was no news at all. Where she was and what she might be doing were a blank.

4. The Stone Crusher

בת

The night sky over the north of England was deepest black, speckled with stars and banded by the mist of the Milky Way, with a bright slice of a first-quarter moon floating in it. The nation was at war and shrouded in blackout, and the heavens had all illumination to themselves.

Edith Kleinmann looked up at the same stars that tracked the skies over Vienna, where her family, God willing, were all keeping safe. She received no news at all, only fears. She longed to know how her mother and father, sister and brothers, friends and relatives were. Edith had news of her own that she was bursting to share. She had met a man. Not just any man, but *the* man. His name was Richard Paltenhoffer, and he was an exile like her.

Her first few months in England had been uneventful. Her work placement, arranged through the Jewish Refugees Committee, the JRC, was as a live-in maid with a Mrs. Rebecca Brostoff, a Jewish lady in her sixties who had a prominent wart on her nose and a home in the quiet suburbs. Her husband, Morris, was a bristle merchant, and they were modestly well off. Both had been born in Russia and had been refugees themselves in their youth.[1]

Leeds was nothing like Vienna; it was a sprawling industrial city, all soot-blackened brick and English Victorian architecture, long streets of small, begrimed factory-workers' houses, grand public buildings and gray, smoky skies. But there were no Nazis here, and although anti-Semitism existed, there was no Jew-baiting, no exclusion, no scrubbing games, no Dachau or Buchenwald.

Many British people were glad to give German Jews a refuge, but some were not, and the government was caught between the two. The press spoke for and against them—emphasizing the contribution they made to the economy and the plight they faced in their home country—while on the other hand British workers worried about their jobs, and their fears were played upon by the right-wing papers. Allegations were made about the criminal tendencies and shiftlessness of Jews, and the threat they posed to the British way of life. But still, there were no actual Nazis, no SA or SS. With the outbreak of war, the government had begun screening foreign nationals and interning enemy aliens; Edith, as a refugee from Nazism, was naturally exempt.[2] And that, it seemed, was that.

Mrs. Brostoff treated Edith—not the world's most natural domestic servant—kindly, and Edith was content on a decent weekly wage of three pounds.

With the country mired in the Phoney War (or the Bore War as some called it), Edith's first winter in England was marked not by conflict but by romance. She had known Richard Paltenhoffer slightly in Vienna; he was the same age, and they had moved in the same circles. In England they met again, and fell in love.

Richard had been through hell since Edith had last seen him. In June 1938, he'd been picked up by the Vienna SS, under the so-called Action Work-Shy Reich. This program was meant to sweep the "asocial" element in German society off the streets and into the concentration camps—the "useless mouths," the unemployed, beggars, drunks, drug addicts, pimps, and petty crooks. Nearly ten thousand people were rounded up this way—many, like Richard Paltenhoffer, just Jews who'd been in the wrong place at the wrong time.[3] Richard had been sent to Dachau, then transferred to Buchenwald,[4] at that time an even worse place than when Fritz and Gustav Kleinmann arrived a year later, more overcrowded and with even more primitive conditions.[5] On one of the regular punishment parades that usually followed evening roll call, a man standing in front of Richard had been bayoneted by an SS guard. The blade passed right through the man, who fell back against Richard, and impaled Richard's leg. The wound gave him trouble for months afterward, but luckily he hadn't succumbed to infection. He was ultimately saved by an extraordinary stroke of luck. In April 1939, to mark Hitler's fiftieth birthday, Himmler agreed to a celebratory mass amnesty of nearly nine thousand concentration camp prisoners.[6] Among them was Richard Paltenhoffer.

Instead of returning to Vienna he crossed the border into Switzerland. The Austrian Boy Scouts organization helped him obtain the necessary travel permit to go to England. By the end of May, he was on his way to Leeds, where he was found a job in a factory making kosher matzohs.[7]

Edith and Richard had both been welcomed into the large and thriving Jewish community in the city, which had its own active branch of the Jewish Refugees Committee. With a tiny budget of 250 pounds a year,* local volunteers helped hundreds find homes and work in Leeds.[8]

It was through a social club for young Jews that Edith and Richard found each other. In Edith's eyes, Richard Paltenhoffer was a reminder of home and the life she had lost—the lively society and her career in fashion rather than in sweeping carpets. Richard was a genial, attractive figure. He had a beaming smile and liked to laugh, and he dressed sharply—nicely cut chalk-stripe suits and a fedora, always with a handkerchief tucked just so in the breast pocket. Among the Yorkshire working men in their serge, woolen scarves, and flat caps, Richard stood out like an exotic bloom in a potato field.

A war—even a phoney war—was a time of possibility for the

* Equivalent to $21,500 in 2019.

young, and with two high spirits far from home it was almost inevitable that they would enjoy themselves to the full. Christmas was past, and January scarcely over when Edith discovered she was pregnant. They started making arrangements for a wedding.

As refugees, any change of status had to be registered with the government. At nine thirty sharp on a Monday morning in February, they presented themselves at the office of Rabbi Arthur Super at the Leeds New Synagogue, and from there they all went to the police station to fill in the required forms. Then, with help from the United Hebrew Congregation, the JRC Control Committee, and a Rabbi Fisher, late of the Stadttempel in Vienna, the prospective marriage was arranged.[9]

With bureaucracy satisfied, on Sunday, March 17, 1940, Edith Kleinmann married Richard Paltenhoffer at the New Synagogue in Chapeltown Road, a remarkable modern building of green copper domes and brick arches in the heart of Leeds' own equivalent of Leopoldstadt.

Two months later, Adolf Hitler launched his invasion of Belgium, the Netherlands, and France. Within a month, the remnant of the British Expeditionary Force had to be evacuated from the beach at Dunkirk. The Phoney War was over. The Germans were on their way, and seemingly unstoppable.

אבא

"*Left*–two–three! *Left*–two–three!"

The kapo barked out the time as the team pulled the quarry wagon up the rails. "*Left*–two–three! *Left*–two–three!" Fritz's shoes slithered on the ice and loose stones, his wasted muscles cracking, hands and shoulders chafed raw by the rope. Around him other men grunted as they pulled. Behind him, still others—his papa among them—pushed with frozen fingers on the bare metal.

Winter had come savagely to the Ettersberg, but the kapos could always be relied on to outdo it. "Pull her, dogs! *Left*–two–three! Onward, pigs! Isn't this fun?" Any man who flagged was kicked and beaten. The wheels squealed and scraped, the men's feet thumped and ground on the stones, their hot breath clouded in the bitter air. "At the double! Faster or you'll be in the shit!"[10] A dozen backbreaking wagonloads to be drawn up this slope to the construction sites every day, each one an hour's round trip. "Forward, pigs! *Left*–two–three!"

"The men-beasts hang in the reins," Gustav wrote, turning his daily hell into a series of stark poetic images. "Panting, groaning, sweating . . . Slaves, cursed to labor, like in the days of the Pharaohs."

There had been a brief respite in the new year; in the middle of January, Dr. Blies, concerned about the extreme death rate from disease in the little camp,[11] and with the SS worrying that it might spread to them, had ordered that the survivors be moved to more sanitary conditions in the main camp. They were showered and deloused, then put into quarantine in a barrack close to the roll-call square. It seemed almost luxurious after the tents, with waxed wooden floors, solid walls, tables to eat off, toilets, and a washroom with cold running water. Everything was kept immaculately clean; the prisoners even had to remove their shoes in an anteroom before entering the barrack. Severe punishments were inflicted for dirt and disorder. During that first blessed week of quarantine they got regular food and didn't have to work. Gustav had regained his strength.

Of course it couldn't last. On January 24, 1940, the quarantine period had ended. For the first time, Gustav and Fritz were separated; Fritz was placed with forty or so other young boys in block 3 (known as the "Youth Block" despite being occupied mostly by grown men).[12]

They got to know the main camp better, learning the layout

and landmarks—the foremost of which was the Goethe Oak. This venerated tree stood near the kitchens and bath block, and was reputed to have been a landmark in Goethe's walks from Weimar up the Ettersberg. So potent were its cultural associations that the SS had preserved it and built the camp around it, putting it to use for punishments.[13] The method, used throughout the concentration camp system, was to tie a man's hands behind his back and hang him by his wrists from a beam or branch. The Goethe Oak made a spectacular venue for this abominable ritual. The hanged men would be left for hours—enough to cripple them for days or weeks—and often beaten bloody while they hung. Two of Gustav's workmates were among those suspended from the Goethe Oak for not working hard enough.

Fritz and his papa had been surprised on emerging from quarantine to learn that Jews made up less than a fifth of Buchenwald's prisoner population.[14] There were criminals, Roma, Poles, Catholic and Lutheran clergy, and homosexuals, but by far the largest number were political prisoners—mostly communists and socialists. Many had been prisoners for years—in some cases since the beginning of the Nazi regime in 1933. However, it was for the Jews and the Roma that the SS reserved their hardest labor and harshest treatment.

"*Left*–two–three! *Left*–two–three!" A dozen loads a day, up the hill, a dozen dangerous high-speed rolls back to the quarry. Fingers burning with cold on the metal, scorched by ropes, minds numbed, feet skittering on the ice, the kapo's abuse.

On and on it went, day after day, until winter began giving way to spring. Eventually Gustav and Fritz were taken off the wagon detail and put to work within the quarry, carrying stones. It almost passed belief that this was even worse.

They had to pick up stones and boulders from where they were hewn out of the rock face and carry them—always at the double—in their bare hands to the waiting wagons. Palms and fingers quickly blistered and bled. The shift lasted ten hours, with a short break at midday. On top of the work came the abuse for which the place was infamous, far beyond anything experienced on the wagons.

"Every day another death," Gustav wrote. "One cannot believe what a man can endure." He could find no ordinary words to describe the living hell of the quarry. Turning to the back pages of his notebook, he began composing a poem—titled "Quarry Kaleidoscope"—translating the chaotic nightmare into precise, measured, orderly stanzas.

> Click-clack, hammer blow,
> Click-clack, day of woe.
> Slave souls, wretched bones,
> At the double, break the stones.[15]

In these lines he managed to find a midpoint between the experiences he lived each day and how it was perceived through the eyes of the kapos and the SS.

> Click-clack, hammer blow,
> Click-clack, day of woe.
> Hear how all these wretches moan,
> Whimpering while they tap on stone.[16]

The slave driving, each endless day, and the murderous abuses all transmuted into poetic imagery. "Shovel! Load it up! Think you can take a breather? You think you're some kind of VIP?" Hands slipping, grazing on the boulders, staining the pale limestone with rust-red blood; struggling, laden, to the wagons. "On, you shirkers—wagon number two! If you don't have it full soon, I'll beat you to a pulp!" The stones clattering and banging into the hollow iron belly of the wagon. "Finished? You think you're free now? D'you see me laughing? Wagon three, at the double! Faster, or you'll be in the shit. On, pigs!" Driven along with kicks and curses; the filled wagon wheeling slowly away up the steep rails: "*Left*–two–three! *Left*–two–three!"

The kapos and guards entertained themselves with the prisoners. One of Gustav's fellow carriers was made to take a huge rock and run with it in circles, uphill and down. "Be funny, understand?" the kapo ordered him. "Or I'll beat you crooked." The victim tried to run in a frolicsome manner, and the kapo laughed and applauded. Round and round he ran, chest heaving, straining for breath, bruised and bloodied. At last, overcome by sheer exhaustion, the droll performance faltered; yet he kept moving, struggling round the circle twice more. But the kapo was bored now; he pushed his victim to the ground and delivered a savage, fatal kick to the head.

A favorite game was to snatch the cap off a passing prisoner and hurl it up a tree or in a puddle—always just beyond the sentry line. "Hey, your cap! Go get it—there by sentry four. Go on, mate, go get it!" This would often be a new prisoner who didn't know the rules. "And the fool runs," Gustav wrote. Past the sentries he would go—*bang!*—and he was dead. Another entry in the guards' escape register, another credit toward some SS man's bonus holiday time: three days for each escapee shot. An SS sentry named Zepp was in cahoots with several kapos, including Johann Herzog, a green-triangle prisoner and former Foreign Legionnaire whom Gustav described as "a murderer of the worst sort."[17] Zepp would reward Herzog and his fellows with tobacco each time they sent a man into the line of his rifle.

Although there were regular suicides, most of the prisoners would not give up and couldn't be tricked. There were some who seemed unbeatable, no matter what abuses were inflicted on them. A blow with a rifle butt:

> Smack!—down on all fours he lies,
> But still the dog just will not die![18]

One day Gustav witnessed a scene which would always remain with him as an image of resistance. In the middle of the quarry, dominating everything, stood a machine. A massive roaring engine drove a series of wheels and belts connected to a huge hopper, into which stones were shoveled. Inside, heavy steel plates worked up and down and side to side, an iron jaw chewing and crushing the stones to gravel. On the footplate, a kapo worked the throttle and gears. When the quarry laborers were not filling the wagons, they were feeding this monstrous machine. For Gustav, the stone crusher was emblematic not just of the quarry but of the camp and the entire system in which Buchenwald was just a component—the great engine in which he and Fritz and their fellows were both the fuel driving it and the grist that it ground.

> It rattles, the crusher, day out and day in,
> It rattles and rattles and breaks up the stone,
> Chews it to gravel and hour by hour
> Eats shovel by shovel in its guzzling maw.
> And those who feed it with toil and with care,
> They know it just eats, but will never be through.
> It first eats the stone and then eats them too.[19]

One prisoner on the crusher-filling detail, a comrade of the man who had been made to run in circles, kept his head down and shoveled the stones, anxious to avoid the attention of the kapos. He was a tall, powerfully built man, and he shoveled well. The kapo on the footplate saw the opportunity for a game; he edged the throttle up until the machine was running at double speed, rattling and banging diabolically. The prisoner shoveled faster. Man and machine labored—the man panting, muscles straining, the crusher grinding and clattering fit to explode. Gustav, working nearby, left off his labors to watch; so did others, and the kapos, also enthralled, let them.

On and on the contest went, shovel by shovel, clattering plates, roaring gears, the man dripping with sweat, the crusher thundering and defecating a cascade of gravel. The man seemed to have tapped within himself an unparalleled seam of strength and will. But the crusher's stamina was limitless, and little by little the man weakened and slowed. Summoning his will, he rallied for one more titanic effort, stretching his muscles to shovel as if for his life; the machine would win, it always won, but still he tried.

Suddenly there was a bang and a long, grinding groan from inside the machine. The stone crusher shuddered, coughed, and stood still. The kapo on the footplate, dismayed, delved into the machine's innards and found that a stone had got into the gears.

There was a silence pregnant with dread. The prisoner leaned on his shovel, gasping for breath. He had vanquished the stone crusher, and was liable to be murdered for it. The senior kapo, stunned for a moment, burst out laughing. "Come here, tall lad!" he called. "What are you, a farmhand? A miner, I'll bet?"

"No," said the prisoner. "I'm a journalist."

The kapo laughed. "A newspaperman? Too bad. I've got no use for one of those." He turned away, then stopped. "Wait, though, I do need someone who can write. Go and wait in the hut there. I have other work for you."

As the hero laid down his shovel, Gustav suddenly felt the weight of the rock in his hands and his kapo's eyes turning toward him. Hurriedly he went back to work, contemplating what he'd just witnessed. Man against machine; on this occasion, man had won a small victory. The machine, it seemed, could be beaten by a person with the necessary strength and will. Whether this was also true of the greater machine remained to be discovered.

The mechanic cleared the stone from the gears and restarted the engine. Rattling, clattering, the crusher went back to work, consuming the rocks fed into its insatiable gullet by the laboring prisoners, eating their strength, their sweat and blood, grinding them down as it ground down the stone.

5. The Road to Life

אמא

Tini regarded the two envelopes with apprehension. They were both identical, from Buchenwald. She knew many wives and mothers whose men had gone to the camps; it sometimes ended with the men obtaining their emigration papers and being released. Or sometimes the men came back to Vienna in a little pot as ashes. Letters were unheard of.

She ripped open one of the envelopes. Inside was something that looked more like an official notice than a letter. Scanning it, she realized with relief that it was from Gustav. She recognized his strong handwriting where he'd filled in his name and prisoner number. Most of the space was taken up by a list of printed restrictions (whether money and packages could be received by the prisoner, whether he could write or receive letters, a warning that enquiries on the prisoner's behalf to the commandant's office would be futile, and so on). There was a tiny space in which Gustav had written a short message, subject to SS censorship. Tini gleaned little other than that he was alive and well and working in the camp. Tearing open the other letter, she found a near-identical message from Fritz. Comparing the two, she noticed that the block numbers were different. So they had been separated. That was a worry. How could the boy look after himself?

Tini's worries were increasing constantly. Since the invasion of France in May, a curfew had been imposed on Jews in Vienna.[1] One might think that there were no further ways in which the Nazis could blight the lives of Jews, but one would

73

be wrong. There was always another stick with which to beat them.

In October the previous year, not long after Gustav and Fritz had been taken away, two transports of Jews had left Vienna bound for Nisko in occupied Poland; they were to be resettled there in some kind of agricultural community.[2] The program stuttered out, but it added to the sense of insecurity among the Jews remaining in Vienna. When the survivors returned home in April, they brought back dreadful stories of abuse and murder.[3]

For Tini, the mission to get her children to safety had become more urgent than ever. With Britain now out of bounds, America was the only hope. Tini's chief concern was to have Fritz released while he was still a minor and eligible for higher priority emigration. She had lodged applications for him, Herta, and Kurt. Each needed two affidavits from friends or relatives living in America, pledging to provide shelter and support. The affidavits were easy enough, as Tini had cousins in New York and New Jersey,[4] and an old and dear friend, Alma Maurer, who had emigrated many years ago and lived in Massachusetts.[5] Support was plentiful—it was the bureaucracy of the Nazi regime and the United States that presented a problem.

President Roosevelt—who wanted to increase the number of refugees taken in—could do nothing against Congress and the press. The United States had a theoretical quota of sixty thousand refugees a year, but chose not to use it. Instead, Washington employed every bureaucratic trick it could dream up to obstruct and delay applications. In June 1940, an internal State Department memo advised its consuls in Europe: "We can delay and effectively stop . . . immigrants into the United States . . . by simply advising our consuls to put every obstacle in the way . . . which would postpone the granting of the visas."[6]

Tini Kleinmann trekked from office to office, queued, wrote letter after letter, filled in forms, suffered the abuse of Gestapo officials, lodged enquiries, and waited and waited and waited, and feared every new message in case it was a summons for deportation. Her every move was blocked by obstacles specifically designed to pander to congressmen and newspaper editors, businessmen, workers, small-town wives and storekeepers in Wisconsin and Pennsylvania, Chicago and New York, who objected stridently to a new wave of immigrants.

Fritz was approaching manhood. Herta was already eighteen and miserably confined without work or opportunity. Ten-year-old Kurt was a worry. Tini fretted all the time about his behavior—he was a good boy, but he had a lot of volatile energy. She worried that he would do something—some trivial act of naughtiness—that would jeopardize all of them.

Keeping her worries to herself, Tini replied to Fritz's and Gustav's brief letters with news of home. She scraped together money to send them, received through charity or earned from occasional illegal work. She wrote that she missed them, and pretended that all was well.[7]

בן

Kurt crept down the staircase to the ground-floor hallway. The street door stood open, and he peeped out. Some boys were playing on the edge of the marketplace—old friends of his from before the Nazis came. He watched them enviously, knowing it was impossible for him to join them.

They had been a happy band, the children from the streets around the Karmelitermarkt. On a Saturday morning his mother would make sandwiches and pack them in his little rucksack. Off he would go with his pals, hiking across the city like a band of pioneers to some distant park, or to the Danube to swim. A perfect society of friends, with no notion that some of them bore a stigma.

Kurt's awareness that some children were not like others

had come on him violently. One day during the first winter, a boy from the Hitler Youth had called him a Jew and pushed him down, shoving his face hard into the snow.

When the hate came from an actual friend—that was when the injustice of it stuck in Kurt's heart. He'd been with a small group of friends in the marketplace—those same boys he was watching now—playing as they had always done. The dominant boy had suddenly decided that he needed to pick on somebody—as such boys will—and singled out Kurt, calling him by the anti-Semitic slurs he'd heard adults use. Then he began pulling off Kurt's coat buttons. Kurt wasn't easily bullied; he hit the boy. Shocked, the boy pulled a metal bar off his little scooter and laid into Kurt with it, battering him so badly that his mother had to take him to the hospital. He remembered her looking down at him as the cuts and bruises on his head were treated. She guessed what would follow. A complaint was made to the police by the boy's parents; Kurt, a Jew, had dared to strike an Aryan. That was a matter for the law. Probably because of his age, Kurt was let off with a caution. After that he understood the malevolence and injustice of this new world.

It was a bewildering place, and the memories it would leave behind were sporadic, vivid impressions.

His mother struggled perpetually to keep him and Herta warm and fed on the little money she could scrape together. There were soup kitchens, and in the summer they went to a farm owned by the IKG to pick peas. There were still a few wealthy Jewish families in Vienna, eking out their remaining money, supporting those who were destitute. Kurt had once gone to dinner with such a family. His mother had coached him strictly: "Sit up straight, behave yourself, do as you're told." Kurt enjoyed a magnificent meal. Except for the Brussels sprouts. He'd never had them before, and hated them, but was too scared not to eat them. He threw up right after.

His social world had shrunk down to his aunts, uncles, and cousins. His favorite was Jenni, his mother's older sister.[8] Jenni had never married; she was a seamstress and lived alone with her cat. She told the children the cat spoke to her: Jenni would ask it a question, and it would say *mm-jaa*. Kurt was never sure whether she was joking. Jenni had a childlike sense of humor and loved animals. She would give him money to buy caps for his pistol so that he could stalk the city pigeon catcher; when the man was about to net some birds, Kurt would fire his pistol, sending them up in a flapping gray cloud, leaving the catcher with an empty net.

Some of Kurt's relatives had married out to non-Jews and now lived in a state of uncertainty, their children classed as *Mischlinge*—mongrels—under Nazi law. One of these cousins was his best friend, Richard Wilczek, whose gentile father had sent him and his mother to the Netherlands for safety after the Anschluss. The Nazis were there too now, and what had become of Richard, Kurt didn't know. Looking out at the street now, it was no longer the same world.

"There you are!" said his mother, and Kurt turned guiltily to face her. "How many times must I tell you not to go outside alone?" Her face was drawn and anxious, and Kurt didn't point out that he hadn't actually stepped outside. "We have to go now. Run and put your coat on."

One of the periodic orders had come down from the Gestapo for all the Jews in the district to report for some inspection or registration or selection. Kurt had picked up on his mother's and Herta's fear, and as the only remaining man in the household, had a plan to protect them. He had a knife. He'd obtained it from another *Mischling* cousin, Viktor Kapelari, who lived in the suburb of Vienna-Döbling. His mother was another of Tini's sisters, who had converted to Christianity when she married. Viktor and his mother were fond of Kurt, and often took him fishing. Mingled with the pleasant memories of these trips, Kurt would always retain a haunting image of Viktor's father the last time he saw him, dressed in the sinister gray uniform of a Nazi officer. After one of their fishing trips, Kurt had come home with a bone-handled hunting knife belonging to Viktor, which he had pocketed.

Putting on his coat while his mother and Herta waited, Kurt slipped the knife into his pocket. The Nazis had taken his father and Fritz away, tormented his sisters, and pushed him down in the snow, beaten him, and made it into *his* crime. There was nothing they would not be allowed to do. He was determined to defend his mother and Herta against them.

Kurt took his mother's hand and they set out for the police station. He fingered the knife blade in his pocket as he walked. He could sense his mother's anxiety, knowing that when Jews were ordered to report, they were sometimes sent away. He guessed that that was what she was afraid of, and felt her distress growing as they approached the police station. To soothe her fears, he showed her the knife.

"See, Mama, I'll protect us."

Tini was appalled. "Get rid of it!" she hissed.

Kurt was astonished and dismayed. "But—"

"Kurt, throw it away before someone sees it!"

There was no convincing her. Reluctantly, he tossed the knife away. They walked on, Kurt almost heartbroken.

As it turned out, the Gestapo didn't do anything bad to them that day. But someday they would. How in the world was he to defend the people he loved now? What would become of them?

כב

Another dawn, another roll call, another day. The prisoners in their stripes stood in ranks in the cool summer air, motionless except to take the sustenance doled out, soundless except to answer to their numbers. Any breach of roll-call discipline meant punishment, as did any infraction of the immaculate neatness and cleanliness of one's barrack block: a veneer of precise order glued over a morass of bestial barbarism.

At last the slow ritual drew to a close. The ranks began to dissolve and re-form into labor gangs. Fritz, looking through the milling crowd, saw his father joining the main quarry detail.

Gustav had had a reprieve during the second half of the winter when Gustav Herzog, one of the younger Jewish block seniors, employed him as bunk room orderly. As an upholsterer he had skill with mattresses, and a knack for keeping things in order. It was illegal, and would have led to punishment for both of them, but it helped the block pass inspections and kept Gustav safe for two months. Eventually, though, the assignment had had to end, and Gustav had been sent back to the murderous task of stone-carrying.

Fritz no longer shared his labor; he'd been transferred to the vegetable gardens attached to the farm complex—hard labor still but infinitely better and safer than the killing ground of the quarry.[9]

Now that they were neither living nor working together, Fritz saw little of his papa, although they met when they could. Money from home enabled them to buy occasional comforts from the prisoner canteen, which helped brighten their days.

As Fritz made his way through the crowd toward his comrades in the garden detail, the camp senior bellowed: "Prisoner 7290 to the main gate, at the double!"

Fritz's heart froze, as if he'd been physically gripped. There were only two reasons for a prisoner to be summoned to the gate at roll call: punishment or assignment to the stone quarry, expressly for the purpose of being murdered.

"Prisoner 7290! Let's have you! Main gate now, at the double!"

Fritz pushed through the mass of prisoners and ran to the gatehouse. Gustav watched him go with his heart in his mouth. Fritz reported to the adjutant, SS-Lieutenant Hermann

Hackmann, a clever, slender young man with a boyish grin that concealed a cynical, brutal nature.[10] He looked Fritz up and down, swinging the hefty bamboo cudgel he always carried. "Wait there," he said. "Face the wall."

He walked off. Fritz stood by the gatehouse in the approved manner, staring at the whitewashed bricks in front of his nose, while the work details marched out. Eventually, when everyone had gone, SS-Sergeant Schramm, Fritz's Blockführer,* came to fetch him. "Come with me."

Schramm led him to the administrative complex that straddled the near end of the Blood Road. Fritz was escorted into the Gestapo building and left standing in a corridor for a long time before being called into a room.

"Cap off," said a Gestapo clerk. "Take off your jacket." Fritz did as he was told. "Put these on."

The clerk handed him a civilian shirt, tie, and jacket. They were rather large for him, especially in his half-starved condition, but he put them on, knotting the tie neatly into the rumpled collar. He was led before a camera, and mugshots were taken from all sides. Utterly unable to imagine any reason for this strange proceeding, Fritz stared with deep, hostile suspicion into the lens, his large, beautiful eyes blazing.

When it was over he was ordered to put his prison uniform back on and run back to the camp. He obeyed, relieved to be in one piece but still with no idea of the purpose of what had just happened. His surprise increased when he was informed that he didn't have to work for the rest of that day.

He sat alone in the empty barrack, wondering. Presumably the clothing had been intended to give the impression that he was living as an ordinary civilian, not as a prisoner, but beyond that he couldn't guess.

---

* SS guard in command of a barrack block.

That evening, when the work details marched back to their blocks, weary and gaunt, Gustav, who'd been in a state of sick anxiety all day, slipped away to Fritz's block. When he looked in through the door and saw him there, alive and well, the relief was immense. Fritz described what had happened, but neither they nor any of their friends could tell what it meant. Anything that involved being singled out by the Gestapo surely couldn't be healthy.

A few days later, it happened again; Fritz was summoned from roll call and taken to the Gestapo office. A copy of his photograph was put in front of him. It was a bizarre image: his face with its shaven scalp and the incongruous suit and tie. If this was meant to give the impression that he lived a normal life, it was preposterous. He was ordered to sign it: *Fritz Israel Kleinmann.*

At last he was told the purpose of it all. His mother had obtained the affidavit she needed from America and had applied for Fritz to be released so that he could emigrate. The photo was required for the application file.

He walked back to the camp on air, feeling hope for the first time in eight months.

בן

"We transferred to the new colony on a fine, warm day. The leaves on the trees had not yet begun to turn, the grass was still green, as if at the height of its second youth, freshened by the first days of autumn."

Stefan's voice filled the room, the only other sound the rustle as he turned a page of the book from which he read.

Fritz and the other boys listened, rapt, to the story of a place that sounded so like and yet so unlike the one in which they lived. Being read to by Stefan was one of the few distractions. Hope still glimmered in the back of Fritz's mind, although it troubled him that the application did not include his papa. Their lives were diverging; Fritz was discovering a wider world through the older prisoners who helped and befriended him.

Foremost among them was Leopold Moses, who had helped Fritz survive in the early months and had remained a friend. Fritz had first encountered him in the quarry, during the dysentery epidemic. Leo had offered Fritz some little black pills: "Swallow them," he said, "they'll prevent the shits." Fritz showed the pills to his father, who recognized them from his wartime service in the trenches; they were veterinary charcoal, and they did help. Leo Moses had taken Fritz under his wing when he was transferred to the Youth Block, and Fritz learned his story. He'd been in the concentration camps since the very beginning. A laborer from Dresden, Leo had been a member of the German Communist Party, and been arrested as soon as the Nazis came to power—long before his Jewishness became an arrestable offense. He'd been briefly a kapo on the haulage column—one of the first Jewish kapos in Buchenwald—but didn't have what it took to be a slave driver; the SS soon demoted him, signing him off with twenty-five lashes on the *Bock*.

Through Leo, Fritz had been befriended by some of the other veteran Jewish prisoners. Here was the key to survival: "It was not good luck; neither was it God's blessing," he recalled later. Rather, it was the kindness of others. "All they saw was the Jewish star on my prison uniform, and that I was a child."[11] He and the other boys often got extra tidbits of food, sometimes medications when they needed them. Among their patriarchs was Gustav Herzog, who had employed Fritz's papa as a room orderly. At thirty-two, Gustl was young for a block senior.[12] The son of a wealthy Viennese family who'd owned an international news agency, he'd been sent to Buchenwald after Kristallnacht. Fritz's greatest respect was reserved for Gustl's deputy, Stefan Heymann.[13] Stefan had the face of an intellectual: high-browed, bespectacled, with a narrow jaw and sensitive mouth. He'd been an officer in the German army in the last war, but as an active communist and a Jew, he'd been among the first arrested in 1933, spending years in Dachau.

On evenings when there was no night work, Stefan would tell stories to take their minds off their plight. This evening he was reading to them from a treasured, forbidden book: *Road to Life* by the Russian author Anton Makarenko. It told the story of Makarenko's work in Soviet rehabilitation colonies for juvenile offenders. As Stefan read, his voice low in the barrack gloom, the boys' camps were brought to life as magical idylls, a universe away from the daily reality of Buchenwald:

> The whispering canopy of the luxuriant treetops of our park spread generously over the Kolomak. There was many a shady mysterious nook here, in which one could bathe, cultivate the society of pixies, go fishing, or, at the lowest, exchange confidences with a congenial spirit. Our principal buildings were ranged along the top of the steep bank, and the ingenious and shameless younger boys could jump right out of the windows into the river, leaving their scanty garments on the window sills.[14]

Most of the boys listening were alone, their fathers having been killed already, and many had grown increasingly apathetic and listless; but hearing this story of another, better world brought them back to life, enthused and cheered.

Other forbidden cultural delights were to be had in Buchenwald. One evening Stefan and Gustl came into the barrack with an air of conspiratorial mystery. Urging Fritz and the other boys to be quiet, they led them across the camp to the clothing store, a long building adjacent to the shower block.

It was quiet and still, the hanging racks and shelves stuffed with uniforms and the clothing confiscated from new prisoners, deadening the echo of the boys' footsteps. Within, some older prisoners had gathered; they gave each boy a piece of bread and some acorn coffee, and then four prisoners appeared with violins and woodwinds. There, in the midst of this musty, cloth-lined room, they played chamber music. For the first time Fritz heard the jaunty, impudent melody of "Eine Kleine Nachtmusik"; the cheerful skipping of the bows on the strings brought the room to life and smiles to the lips of the prisoners gathered round. It was a memory Fritz would treasure: "For a very short time we were able to laugh again."[15]

Outside these few borrowed hours, there was no laughter.

Working in the vegetable gardens, whose produce was sold in Weimar market or to prisoners in the canteen, was an improvement on the quarry, but tougher than the boys had expected. They had anticipated being able to pilfer a few of the carrots, tomatoes, and peppers they planted, but they were never allowed near the ripened crops.

The gardens were under the overall authority of an Austrian officer, SS-Lieutenant Dumböck. Having spent time in exile with the Austrian Legion when the Nazi Party was outlawed, Dumböck now dedicated himself to persecuting Austrian Jews in revenge. "You pigs ought to be annihilated," he told them repeatedly, and did his best to make it come true. He was reckoned to have murdered forty prisoners with his own hands.[16]

Fritz was assigned to *Scheissetragen*—shit-carrying.[17] They had to collect the liquid slop of feces from the prisoner latrines and the sewage plant and carry it in buckets to the vegetable beds. Every trip, there and back, had to be done at top speed, running as fast as one could with the noisome, glooping pails of filth. The only job worse than shit-carrying was the so-called "4711" detail, named after the popular German eau de cologne; their job was to scoop out the feces from the latrines—often with their bare hands—to fill the shit-carriers' buckets. The SS typically allotted this task to Jewish intellectuals and artists.[18]

At least the boys were treated decently by their kapo, Willi Kurz. A former amateur heavyweight boxing champion, Willi was a disillusioned soul; having once been on the board of an Aryans-only sports club in Vienna, he had been deeply hurt when the authorities looked into his ancestry and branded him a Jew.

He was kind to the boys on his detail; he let them ease off the pace and take a rest if the SS weren't around. Whenever a guard appeared, Willi would make a show of driving the boys along at full pace, yelling savagely at them and brandishing his cudgel, but he never used it on them. His performance was so convincing that the guards didn't bother inflicting beatings themselves if Willi was in charge.

All the while, Fritz thought back on the photograph and hoped.

אבא

"*Left*–two–three! *Left*–two–three!"

Gustav, with his shoulder to the rope, heaved. There was no pause, no respite—just heave, step, heave, step, into eternity. On either side the other cattle heaved and stepped, sweating in the sunlight dappled by the trees. Twenty-six Jewish stars, twenty-six half-starved bodies hauling the wagon with its load of logs through the forest, up the slope, along the dirt road, wagon wheels groaning under the weight.

It was arduous, but for Gustav, transfer out of the quarry to the haulage column had been a lifesaver, and he owed it to Leo Moses. The quarry had become worse than ever. Prisoners were chased over the sentry line every day, and Sergeant Hinkelmann had invented a new torture: if a man collapsed from exhaustion, Hinkelmann would have water poured into his mouth until he choked. Meanwhile, Sergeant Blank entertained himself by throwing rocks down on the prisoners as they left the quarry; many were hit and maimed, and some killed. The

SS men had also taken up an extortion racket against Jewish quarry workers who received money from home; every few days, each man had to pay up five marks and six cigarettes or get beaten. With two hundred prisoners, the guards made a tidy little income from their "paydays," although the sum declined week by week as the prisoners were murdered.

Through Leo's advocacy, in July, Gustav had been transferred out of the killing ground to the haulage column. They carried building materials around the camp complex all day long—logs from the forest, gravel from the quarry, cement from the stores. The kapos made them sing as they worked, and the other prisoners called them *singende Pferde*—singing horses.[19]

"*Left*–two–three! *Left*–two–three! *Left*–two–three! Sing, pigs!"

Whenever they passed an SS guard, he would lash out at them. "Why don't you run, you dogs? Faster!"

But still it was better than the quarry. "It is hard work," Gustav wrote, "but one has more peace and is not hunted . . . Man is a creature of habit, and can get used to everything. So it goes, day after day."

The wheels turned, the men-horses sang and heaved, the kapos yelled time, and the days passed.

ב

SS-Sergeant Schmidt screamed at the group of men as they ran in circles around the roll-call square. "Faster, you Jew-shits!" Fritz and the other boys, who were at the front, increased their pace to avoid the blows Schmidt aimed at anyone going too slowly. Some of the runners struggled with painful stomachs or testicles where Schmidt had kicked them for answering too slowly at roll call. "Run! Run, you pigs, run! Faster, you shits!"

While the other prisoners returned to their barracks, the block 3 inmates had been made to stay. Schmidt, their Blockführer, had found fault in his inspection again—a bed not properly made, a floor insufficiently pristine, belongings not stowed away—and it was punishment time once more: *Strafsport*. Thickset and flabby, Schmidt was a noted goldbrick as well as a sadist; he held a post in the prisoners' canteen and skimmed off tobacco and cigarettes in large quantities. The boys of block 3 called him "Shit Schmidt" after his favorite word.[20] "Run! March! Lie down...get up... That's shit—lie down again. Now run!" *Thwack* went his bullwhip on the backside of some poor man who couldn't keep up. "Run!"

Two hours went by, the hot sun setting and the square cooling, the men and boys sweating and laboring for breath. At last Schmidt dismissed them with a curse and they limped back to their block.

Starving, they sat down to the only warm meal of the day: turnip soup. If they were lucky there might be a little scrap of meat.

Fritz had finished and was about to get up when Gustl Herzog told the boys to stay where they were. "I have to talk to you," he said. "You boys mustn't run so fast during the *Strafsport*. When you run fast, your fathers can't keep up and they get beaten by Schmidt for lagging behind." The boys were ashamed, but what could they do? *Someone* would get beaten for going too slow. Gustl and Stefan showed them the solution. "Run like this—lift your knees higher, take smaller steps. That way, you look like you're running flat-out, but you go slowly."

It worked well enough to fool Shit Schmidt. As time went on, Fritz would learn all the veterans' little tricks—absurd things, some of them, but they could mean the difference between safety and pain, or between life and death.

And all the while, as Fritz labored in the gardens and Gustav hauled his wagon, the war went on in the world outside, the months dragged by, and all hope of release slowly waned. His mother's application to have him released, which had sustained Fritz's spirits for a while, began to fade into the realm of no hope.

6. A Favorable Decision

בת

Everything was changing for Edith and Richard. In this land of refuge, they now saw something emerging that they'd thought was left behind in Vienna.

In June 1940, the quiet home front became a place of bombs and blood and death, the Bore War giving way to the Battle of Britain. Every day, Luftwaffe bombers attacked airfields and factories, and every day the Spitfires and Hurricanes scrambled to oppose them. The RAF had become a coalition force, its British and Commonwealth pilots joined by exiles from Poland, France, Belgium, and Czechoslovakia. Britain still liked to think of itself as a sole nation, but it was nothing of the sort.

The press fixated on two things: the progress of the battle and growing fears about Britain being infiltrated by German spies and saboteurs paving the way for invasion. The rumors had begun in April. The press—with the *Daily Mail* at the forefront—had helped whip up paranoia about fifth columnists.[1] Paranoia became hysteria, and hostile eyes were turned toward the 55,000 Austrian and German Jewish refugees; these men, women, and children were hardly likely to be spying for Hitler, and had been spared internment,[2] but with the country under threat of invasion, the *Mail* and some politicians were strident in their demands that the government intern all German nationals, regardless of status, for the sake of national security.

When Churchill became prime minister in May, he extended

the categories subject to internment to include members of the British Union of Fascists, the Communist Party, and Irish and Welsh nationalists. In June, he lost patience and issued the order: "Collar the lot!"[3] To avoid putting too much pressure on infrastructure, arrests would proceed in stages. Stage one: Germans and Austrians—Jews, non-Jews, and anti-Nazis alike—who did not have refugee status or who were unemployed. Stage two would sweep up all remaining Germans and Austrians living outside London, and the third stage would round up those in London.

Churchill told Parliament, "I know there are a great many people affected by the orders . . . who are the passionate enemies of Nazi Germany. I am very sorry for them, but we cannot . . . draw all the distinctions which we should like to do."[4] The first-stage arrests began on June 24.[5]

People were circulating the kind of anti-Semitic slurs that always arose in times of stress: Jews were black marketeers, evaded military service, enjoyed special privileges, had more money, better food, better clothes.[6] Desperate to quell the growing anti-Semitism, the Anglo-Jewish community fell into line with the national mood. The *Jewish Chronicle* breathtakingly recommended taking "the most rigorous steps" against refugees, including Jews, and supported the extension of internment; British synagogues stopped allowing sermons in German, and the Board of Deputies of British Jews began to restrict gatherings of German–Jewish refugees.[7]

In Leeds, Edith's fears had been growing for months. She and Richard had set up home in an apartment in a rather rundown Victorian house close to the synagogue.[8] Edith had left her live-in position with Mrs. Brostoff and switched to a daily job as a cleaner for a woman who lived nearby. This was no light undertaking, as changes of employment by refugees had to be registered and approved by the Home Office.[9] Richard continued with his kosher matzoh-baking. With a baby on the way, they should have been happy, but Edith was deeply unsettled. Life was now uncomfortable for anyone with a German accent in Britain. And with a German invasion looking certain, they were consumed by fear. They had seen how quickly Austria had fallen to the Nazis, and it was only too easy to imagine storm troopers in Chapeltown Road, and Eichmann or some other SS ghoul issuing orders from Leeds Town Hall.

Feeling that the time had come to try to get out of Europe altogether, Edith dug out the affidavits from her relatives in America. She enquired of the Refugees Committee whether they were still valid now that she was married. It took nearly two weeks for the reply to find its way back from London: no, they were not. Edith would need to write to her sponsors and ask them for new affidavits. The sponsors would also need to extend them to include her husband.[10] And of course they would need to apply for an immigration visa at the US Embassy in London. With the war escalating in the skies over their heads and the threat of internment growing, Edith and Richard were looking at an excruciatingly long process.

They would never discover just how long it would take; at the beginning of July the second stage of the government program came into force, and the Leeds police arrested Richard.

It was only by luck that they didn't take Edith too. Women with children were not exempt, but pregnant women were.[11]

Richard was only twenty-one years old, with the scars of Dachau and Buchenwald on his body; he had fled to Britain seeking sanctuary. And now he had been torn from his wife and unborn child and imprisoned by the very people who should be shielding him from the Nazis.

Edith immediately lodged an application for his release with the Home Office. It was no easy process, as internees had to prove that they were no threat to security and that they could make a positive contribution to the war effort.[12] Both the Leeds and London branches of the Jewish Refugees Committee lobbied the Home Office on behalf of the thousands now incarcerated. Many weren't even in real camps with proper facilities—the numbers had been too great, and improvised centers had been set up in derelict cotton mills, old factories, racetracks, and anywhere that could be found. Many went to the main internment center on the Isle of Man.[13] Some internees were old enough to remember that the Nazi concentration camps had begun precisely like this—Dachau had been founded in a derelict factory.

The weeks of July and August went by, Edith's pregnancy advanced, and no word came. She wrote to the JRC in late August, but they advised her against pressing the matter: "We . . . feel that you have done everything possible at the present time and think it would be most unwise for our Committee to intervene. We have been advised by the Home Office that additional appeals and letters of enquiry . . . can result in delaying any decision."[14]

A few days later, the decision was reached—Richard would remain in custody.

For a concentration camp veteran, life in an internment camp was relatively mild. There was no forced labor, no real punishments, no sadistic guards. Internees played sports, set up newspapers, concerts, and educational circles. But they were prisoners nonetheless. And although there were no SS guards, Jews often found themselves confined alongside unrepentant and vindictive Nazi sympathizers. Richard had the additional torment of knowing that Edith was having to cope with her pregnancy alone and without his wages.

In early September, now in her ninth month, Edith submitted a second application for Richard's release. The JRC assured her, "We sincerely trust that the application will receive a favorable decision."[15] The waiting began again. After two weeks, a brief note came from the Aliens Department of the Home Office, telling her that Richard's case would come before the committee "as soon as possible."[16]

Two days later Edith's contractions started. She was taken to the maternity hospital in Hyde Terrace in the center of Leeds, where on Wednesday, September 18, she gave birth to a healthy baby boy. She named him Peter John. An English name for a Yorkshire-born English baby son.

As that fraught summer drew to a close and the public spirit grew phlegmatic, the mood turned against interning harmless refugees. In July, a ship carrying several thousand internees to Canada—including some Jews—had been sunk by a U-boat. The loss of life had made Britain look at itself and take stock of how it was treating innocent people just because they were foreign. The policy was gradually reversed. In Parliament, politicians expressed regret for what they had done in a fit of panic; one Conservative member said, "We have, unwittingly I know, added to the sum total of misery caused by this war, and by doing so we have not in any way added to the efficiency of our war effort."[17] A Labor member added: "We remember the horror that sprang up in this country when Hitler put Jews, socialists, and communists into concentration camps. We were horrified at that, but somehow or other we almost took it for granted when we did the same thing to the same people."[18]

Peter was five days old when the news came through to Edith—Richard had been released.[19]

אבא

Gustav opened his notebook and leafed through the pages. So few of them—the whole of 1940 summed up in just three pages crammed with his strong script. "Thus the time passes," he wrote, "up early in the morning, home late in the evening, eating, and then straight to sleep. So a year goes by, with work and punishment."

It wasn't always straight to sleep. A new torment had been devised for the Jews by the deputy commandant in charge of the main camp, SS-Major Arthur Rödl. Each evening, when they returned from the quarry, the gardens, and the building sites, exhausted and hungry, while all the other prisoners went to their barracks, the Jews were made to stand on the roll-call square under the glare of the floodlights and sing.

Rödl, a bumptious crook whose low intelligence had not hindered his rise to senior rank, loved to hear his Jewish "choir" perform. The camp orchestra sat to one side playing, and the "choirmaster" stood on top of a gravel heap at the edge of the square and conducted.

"Another number!" Rödl would call out over the loudspeaker, and the weary prisoners would draw breath and struggle through another song. If the singing wasn't good enough, the loudspeaker would bark out: "Open your mouths! Don't you pigs want to sing? Lie down, the whole rabble, and give us a song!" And down they had to lie, whatever the weather, in the dust, the dirt, muddy puddles, or snow, and sing. Blockführers would walk between the rows, kicking any man who didn't give voice loudly enough.

The ordeal often went on for hours. Sometimes Rödl would grow bored and announce that he was going for dinner, but the prisoners would have to stay and practice: "If you can't get it right," he told them, "you can stay and sing all night." The SS guards, who resented having to stand by and supervise, would take out their wrath on the prisoners, administering kickings and beatings.

They sang the "Buchenwald Song" more often than anything else. Composed by the Viennese songwriter Hermann Leopoldi with words by celebrated lyricist Fritz Löhner-Beda, both of whom were prisoners, it was a stirring march tune, with words extolling courage in the midst of wretchedness. It had been specially commissioned by Rödl: "All other camps have a song. We must get a Buchenwald song."[20] He'd offered a prize of ten marks to the successful composer (which was never paid) and was delighted by the result. The prisoners sang it when they marched out to work in the mornings:

O Buchenwald, I cannot forget you,
For you are my fate.
He who has left you, he alone can measure
How wonderful freedom is!
O Buchenwald, we do not whine and moan,
And whatever our fate may be,
We will say yes to life,
For the day will come when we are free!

Rödl abjectly failed to recognize the spirit of defiance. "In his weakness of intellect," recalled Leopoldi, "he absolutely did not see how revolutionary the song actually was."[21] Rödl also commissioned a "Jewish Song" with defamatory lyrics about the crimes and pestilence of the Jews, but it had been "too stupid" even for him, and he banned it. Some other officers later resurrected it and forced the prisoners to sing it late into the night.[22]

But still it was the "Buchenwald Song" more often than not. The Jews sang it countless times on the roll-call square under the lights; "Rödl enjoyed dancing to the melody," said Leopoldi, "as on one side the camp orchestra played, and on the other side the people were whipped."[23] Singing it on the march to work in the red dawn, they invested it with all their loathing and hatred of the SS. Many died singing it.

"They cannot grind us down like this," Gustav wrote in his diary. "The war goes on."

בן

Buchenwald expanded month by month. The forest was eaten away and logged, and amid the waste the buildings rose like pale fungus on the blighted back of the Ettersberg.

The SS barracks gradually formed a semicircle of two-story blocks, with an officers' casino in the center. There were handsomely designed villas with gardens for the officers, a small zoo, riding and stable facilities, garage complexes, and a fuel station for SS vehicles. There was even a falconry establishment. Standing among the trees on the slope near the quarry, it comprised an aviary, a gazebo, and a Teutonic hunting hall of carved oak timbers and great fireplaces, stuffed with trophies and heavy furniture. It was intended for the personal use of Hermann Goering, but he never even visited the place. The SS was so proud of it that for a fee of one mark local Germans could come in and look around.[24]

All of this construction was wrought from the rock and trees of the hill on which it stood, and mixed with the blood of the prisoners whose hands transported and laid the stones and bricks and timbers.

Along the roads between the construction sites, Gustav Kleinmann and his fellow slaves hauled their wagons of materials, and his son was now one of those whose hands put up the buildings. Fritz's tireless benefactor, Leo Moses, had again used his influence to have Fritz transferred to the detail responsible for building the SS garages.[25]

The kapo of Construction Detachment I, which was undertaking the project, was Robert Siewert, a friend of Leo Moses. A German of Polish extraction, Siewert wore the red triangle of a political prisoner. He'd been a bricklayer in his youth, and served in the German army in the last war. A dedicated communist, he'd been a member of Saxony's parliament in the

1920s. In his fifties, he had an air of resilient strength and energy: thickset, with a broad face and narrow eyes under dark, heavy brows.

At first the labor was all about carrying—bring this here, bear this burden, and *run*! A one-hundred and ten pound sack of cement weighed more than Fritz himself. Laborers in the yard would lift it onto his shoulders, and he would carry it, staggering, trying to run, to wherever it was needed. But there was no abuse, no beatings. The SS valued the construction detachment highly, and Siewert was able to protect his workers.

For all his stern appearance, Robert Siewert had a kind heart. He reassigned Fritz to the lighter task of mixing mortar, and taught him how to gain favor with the SS. "You have to work with your eyes," he told him. "If you see an SS man coming, work fast. But if no SS are about, then you take your time, you spare yourself." Fritz became so adept at watching for the guards and making a show of intense productive labor that he acquired a reputation for industriousness. Siewert would point him out to the construction leader, SS-Sergeant Becker, and say, "Look how diligently this Jewish lad works."

One day Becker arrived at the building site with his superior, SS-Lieutenant Max Schobert, deputy commandant in charge of protective custody prisoners. Siewert called Fritz away from his work and presented him to the officer, extolling his performance. "We could train Jewish prisoners as bricklayers," he suggested. Schobert, a brutal-faced individual with a perpetual sneer, looked down his large nose at Fritz. He didn't like this suggestion at all; all that expense to train Jews! Oh no, he wouldn't permit that. Nonetheless, a seed had been planted.

When new SS troops arrived at Buchenwald to bring the garrison up to full strength, the seed began to germinate. Work had to be accelerated to complete the SS barracks—a task well beyond the capacity of the existing workforce. Siewert pressed his case again, this time taking it all the way to Commandant Koch. He complained that he didn't have enough bricklayers. The only solution would be to train young Jews for the job. Koch's reaction was the same as Schobert's. Siewert insisted that he simply couldn't provide the labor in any other way, but the answer remained *no Jews*.

Siewert decided there was nothing for it but to prove his case. Fritz became his apprentice. Siewert began by having him taught to lay bricks to build a simple wall under the supervision of Aryan workmen. With a string laid out as a guide, he pasted on the mortar and laid down brick after brick, neatly and correctly. Fritz had inherited his father's aptitude for manual craft, and he learned quickly. Having mastered the basics, he was taught how to do corners, pillars, and buttresses, then lintels, fireplaces and chimneys. In wet weather he learned plastering. Every day Siewert would come and talk to him and check on his progress. In double-quick time Fritz became a quite passable mason and builder—the first Jew in Buchenwald to do so.

His progress was so impressive, and the need so urgent, that Commandant Koch relented, allowing Siewert to start up a training program for Jewish, Polish, and Roma boys. They spent half of each day working onsite, and half in their block in the camp being taught construction theory and science. They wore green bands on their sleeves with the inscription "Brick-layers' School" and enjoyed certain privileges. A particular delight was the heavy laborers' special food allowance: twice a week, they received an extra ration of bread and a pound of blood pudding or meat pâté, which was brought to the construction site for them. This was on top of their standard daily ration of bread, margarine, a spoonful of curd or beet jam, acorn coffee, and cabbage or turnip soup.

To Fritz, Robert Siewert was a hero, representing the spirit of resistance and the essence of human kindness. The young were his greatest concern, and he did all he could to equip them with skills and knowledge that could save their lives. "He spoke to us like a father," Fritz would recall, "with patience and kindness."[26] Fritz wondered where he got the strength, at his age and after so many years of imprisonment.

When winter began to set in, Siewert got permission to set up oil drums as braziers on the building site, on the pretext that the plaster and mortar were liable to crack in freezing conditions. His real purpose was the welfare of his workers, who had only their thin prison uniforms to protect them. A humane and courageous man from heart to backbone, Robert Siewert never failed in his duty, knowingly putting himself at risk by interceding with the SS on behalf of Jews, Roma, and Poles.

But Siewert's influence did not extend far beyond the limits of the construction site and the bricklaying school. As soon as the day's work ended and the prisoners returned to the main camp, it was back to the regime of singing parades, random beatings, food deprivation, and capricious murders. Fritz would look at his fellow prisoners and silently give thanks that at least he ate better than they did and was not at risk of being driven over the sentry line or kicked to death. He ached for his papa, who slaved each day on the haulage column. Fritz saved what he could from his additional rations to give to him when they met in the evenings.

Gustav's mind was eased by his son's new status and the safety it brought. "The boy is popular with all the foremen and kapo Robert Siewert," he wrote. "From Leo Moses we get our greatest support, which gives us further confidence." To Gustav's indomitably optimistic mind, it was beginning to seem like they might survive this ordeal.

Fritz had been moved out of the Youth Block earlier in the year and transferred to block 17, close to his father's block. It had been painful to part from his friends, but the move proved formative, another important stage in his growth to manhood.

Block 17 was where the Austrian VIPs and celebrity prisoners—the *Prominenten*—were housed.

Most were politicals, but of a higher status than most of the red-triangle men in the camp.[27] Some of their names were familiar to Fritz, his father having known of them in his time as an activist in the Social Democratic Party. The inmates included Robert Danneberg, a Jewish socialist who had been president of the Vienna provincial council and one of the leading figures in "Red Vienna"—the socialist heyday that had lasted from the end of the First World War until the right-wing takeover in 1934. Contrasting with Danneberg's sober presence was the droll, round-faced Fritz Grünbaum, star of the Berlin and Vienna cabaret scenes, *conférencier,** scriptwriter, movie actor, and librettist for Franz Lehár (one of Hitler's favorite composers). As a prominent Jew and a political satirist, Grünbaum had been taken by the Nazis soon after the Anschluss. Aging and slightly built, with his bald, shaved pate and bottle-bottom spectacles, he resembled Mahatma Gandhi. Having endured both the quarry and latrine details, his health and spirit had been broken, and he had attempted suicide. Even so, he managed to keep up a semblance of his old persona, and would on occasion perform cabaret for the other prisoners. His comment on his plight as a Jew was simple and to the point: "What does my intellect benefit me when my name damages me? A poet called Grünbaum is done for." He was right; he would be dead within months.[28]

Fritz also got to know the bespectacled, somber-looking Fritz Löhner-Beda, author of the poignant, defiant lyrics of the "Buchenwald Song." Like Grünbaum, he had written librettos for Lehár's operas. He always hoped that Lehár, who had influence with both Hitler and Goebbels, would be able to have him freed, but he hoped in vain. To add to his torment, songs

* Cabaret emcee.

from Lehár's operettas *Giuditta* and *The Land of Smiles* were often played over the camp's loudspeaker, the SS apparently unaware of his involvement in them. Even more hurtfully they played the popular song "I Lost My Heart in Heidelberg," for which he had written the lyrics.

One of the brightest of the block 17 *Prominenten* was Ernst Federn, a young Viennese psychoanalyst and Trotskyist who wore the red-on-yellow star of a Jewish political prisoner. Forbidding to look at, with heavy features that made him look almost thuggish beneath his cropped scalp, Ernst was the kindest of souls. Anyone could come to him to talk about their problems. His irrepressible optimism earned him a reputation for being a little crazy, but gave marvelous encouragement to the other prisoners.[29]

There were many social democrats, Christian socialists, Trotskyists, and communists in block 17. In the free time in the evenings, young Fritz would sit and listen to their conversations about politics, philosophy, the war . . . Their talk was intellectual, sophisticated, and Fritz strained to comprehend what they said. One thing that came through clearly was the strength of their belief in the idea of Austria. Despite their own hopeless situation and their country's obliteration as an independent state, they shared a vision of a future Austria, free from Nazi rule, renewed and beautiful. The men of block 17 were convinced that Germany would lose the war in the end, even though the trickles of news reaching the camp indicated that they were currently winning on all fronts.

Fritz's faith and courage grew in the light of these men's vision of a better future, even while guessing that few of them would live to see it. "The camaraderie I learned in block 17 changed my life fundamentally," he would recall. "I became acquainted with a form of solidarity unimaginable in life outside the concentration camps."[30]

A highlight of Fritz's time in the block was Fritz Grünbaum's birthday celebration, it being the same day as his sister Herta's (she would be turning eighteen that day). The block 17 men saved portions of their rations to give the old fellow a decent dinner, and a little extra was stolen from the kitchens. After their meal, Löhner-Beda gave a speech and Grünbaum himself sang a few verses. As the youngest inmate present, Fritz was permitted to congratulate the humbled star.

With these politicians, intellectuals, and entertainers, what could one young apprentice upholsterer-turned-bricklayer from Leopoldstadt, a playmate of the Karmelitermarkt, possibly have in common? That they were all Austrians either by birth or by choice, and that they were Jews. It was enough. In Buchenwald, they were a tiny nation of survivors surrounded by a poison sea.

And the deaths went on.

The killings in the quarry were growing more frequent. Many of the dead were friends of Fritz's or his papa's, some from the old days in Vienna. That year, across all the concentration camps, prisoner deaths spiraled from around 1,300 to 14,000.[31] The atmosphere of war was the cause of it; while the Waffen-SS and the Wehrmacht fought and conquered Germany's enemies from Poland to the English Channel, the Totenkopf SS in the camps felt their blood stirring and their tempers flaring, and they ramped up their war against the enemy within. News of military victories triggered spurts of triumphal aggression, and setbacks—such as the failure to subdue Britain, the only enemy still fighting—inspired retribution.

Disposal of the rising numbers of corpses became a problem, and in 1940 the SS began to equip its camps with crematoria.[32] Buchenwald's was a small, square building with a yard surrounded by a high wall. From the roll-call square the spike of its chimney could be seen under construction, brick upon brick; when it was complete, it began pouring out its first acrid smoke. From that day on, the smoke would scarcely stop. Sometimes it blew away across the treetops; often it drifted over the camp. But always there was the smell of it: the bitter odor of death.

אמא

In the new year, after months of frustration, Tini received a result at last from the US consulate in Vienna.

Since March 1940, there had been a standing summons for an interview for emigration, but Tini had been advised that she needed to wait until Gustav and Fritz had been freed if she wanted the family to go together.[33] But since the SS would not release prisoners until they had all the necessary papers to emigrate, this was a hopeless dead end.

All the affidavits were in place. The problem was getting American visas and valid tickets for travel (which had to be paid for) *and* having everything coordinated. While France remained free, it had provided a route out of Europe to America, but the German invasion had closed the French ports. In the autumn, Lisbon had become available to emigrants, but the US consulate in Vienna simultaneously put a freeze on issuing visas. Roosevelt's stance on giving a haven to refugees had withered in the face of America's growing anti-Semitism. Capitulating to public opinion, the president had instructed the State Department to reduce the number of visas to near zero: "No more aliens." The consulate still called applicants for interview, which was tortuous in itself, requiring expenditure on notarized documents, police certificates, steamship tickets, local anti-Jewish taxes. At the final interview, when the anxious applicant had miraculously got every document in order, they were told that they had failed to show they could make a contribution to the United States, and that they were therefore likely to "become a public charge."[34] *Visa refused.*

As of October 1940, virtually all applicants—people who were living in constant terror and had beggared themselves satisfying the bureaucratic requirements—went away heart-broken.[35] Tini was close to despair. "We have everything," she wrote to the German Jewish Aid Committee in New York, "but none of us has emigrated . . . Our local consulate is not giving us adequate answers."[36] She couldn't understand the endless frustration; her husband was a hard worker with good skills, and they had affidavits in plenty.

Her only hope was for the children. At the beginning of 1941, Tini made her breakthrough. Her old friend Alma Maurer, who had been at her wedding and now lived in Massachusetts, had obtained an affidavit for Kurt from a prominent Jewish gentle-man in the town where she lived—a judge no less. And then a miracle—the United States was willing to make an allowance for a small number of Jewish children. In conjunction with the German Jewish Children's Aid organization, a limited number of unaccompanied minors would be received and placed with appropriate Jewish families in the United States. Kurt had been accepted.

It would hurt both Tini and Herta to let him go, but it was the only way to get him to safety. And there was more good news—although Herta wasn't eligible for the children's scheme, the kind gentleman in Massachusetts would be willing to spon-sor her if she could obtain the necessary visa.

7. The New World

אבא

Beneath a cloud-packed gray sky the Ettersberg lay under thick snow which softened but did not hide the outlines of the barrack blocks and the tower-spiked fences.

Gustav leaned on his shovel. The kapo's back was turned, and Gustav snatched the moment to catch his breath. His bare hands were purple and when he breathed on them there was no feeling of warmth—no sensation at all. He knew that when he returned to the barrack in the evening and the bone-cold numbness leached out of him, they would gripe and ache abominably.

A new year, but nothing changed in this world except the passing of seasons and the daily passing of lives. Smoke from the crematorium drifted in the freezing air, bringing into the prisoners' nostrils the scent of their own futures.

Gustav sensed the kapo turning toward him, and was already plying his shovel before the man's eyes reached him. The work of the haulage column had been interrupted by the snow; each day the team shoveled the camp streets clear, hauled the snow away, and each night nature buried them deep once more.

The light was fading. With no eyes upon him, Gustav rested again. He looked up at the southeastern sky, marbled gray and speckled with falling flakes, smeared with smoke. Somewhere over there, far beyond these fences and woods, was his home, his wife, Herta, little Kurt. What were they doing right now? Were they safe? Warm or cold? Frightened or hopeful?

Despairing? He and Fritz still received letters from Tini, but it was no substitute for being there.

With a last glance at the sky, Gustav bent his back and drove his shovel into the snow.

בן

The sky above Kurt's head was warm and blue, shimmering with the sunlight-dappled leaves of horse chestnuts studded with snowy blossom. He put one foot before the other, gazing upward, dizzying himself with pleasure.

Looking ahead, he realized that he had lagged behind the rest of the family. His mama and papa were walking arm in arm, Fritz sauntering with his hands in his pockets, Herta strolling prettily, Edith upright and elegant.

They had spent the morning in the Prater, and Kurt was replete with delight. He'd lost count of the number of times he'd shot down the great slide—if you helped carry bundles of mats back to the top, the man in charge gave you a free ride, and Kurt and Fritz and the other less well-off kids always took a few turns. Now, strolling along the Hauptallee, the broad avenue that ran through the Prater woods, Kurt was amusing himself by walking with one foot on the path and the other on the raised grass bank between it and the road. His senses full, he didn't notice that the rest of the family were getting farther and farther ahead. He hummed to himself, enjoying the sensation of boosting himself up on each high step. All awareness of time slipped away, and when he looked up again, he was alone.

An instant's terror flickered in his chest. Before him, the ranks of trees receding into the distance, the woods on either side, the families, the couples, bicycles and carriages and cars swishing by on the road; through the trees, the colors of the amusement park and more people—but nowhere could he pick out the familiar shapes of his parents or his sisters or Fritz. They had simply vanished, as if snatched away.

The momentary terror passed. There was no need to panic; Kurt knew the Prater like he knew the face of a friend; it was little more than a half mile from home. He could find his own way. The Hauptallee opened out on to the Praterstern, a huge star-shaped roundabout where seven boulevards and avenues met. After the peace of the woods, it was a maelstrom of noise and movement; trucks, motor cars, and trams streamed roaring from left to right, pouring out of the nearest boulevards on to the roundabout; the pavements teemed with people.

Kurt realized that he had no idea what to do now. He had come through this place times without count, but always with a grown-up or older sibling. He'd never needed to pay attention to how you got through this torrent.

After a while he became aware of a woman's voice. He glanced up and found a lady peering down at him with concern. "Are you lost?" she asked. Well, he wasn't lost; he knew the way but couldn't figure out how to physically accomplish it. He also didn't know quite how to explain this complex concept. The lady frowned anxiously at him.

A policeman appeared and quickly took control. He took Kurt by the hand and led him back toward the Prater, bearing left along Ausstellungsstrasse. Eventually they came to the police station, a large and extremely important-looking building of red brick and ashlar. Kurt was led into a world of dark uniforms and quietly efficient bustle filled with strange smells and sounds. He was given a seat in an office. A policeman working there smiled at him, chatted, played with him. Kurt had a roll of caps, and to his delight the policeman, using the buckle of his dress belt, set them off one at a time, the banging echoing round the office like rifle fire. Distracted and enjoying the policeman's company, Kurt scarcely noticed the time passing.

"Kurtl!" He spun around at the sound of the familiar voice. "There you are!" There was his mama in the doorway, and his papa behind her. His heart lit up; he jumped up and ran toward his mother's open arms.

בֵּן

Kurt woke, staring and shaking, his heart pounding. For a moment he had no idea where he was. A rush of sound, thudding, clattering in his ears; beneath him a hard wooden bench; around him strange people; a sensation of rocking rhythmically. He noticed the thin wallet lying on his chest, and remembered.[1]

This was the train to his new life.

The slatted wooden bench had numbed his backside but he'd been so tired, sleep had taken hold of him, and he'd slumped against the passenger beside him. He sat up and touched the wallet. He recalled his mother hanging it round his neck.

That image was vivid: they were in the kitchen in the apartment. She sat him on the table—the same worn surface where he had once helped her roll up the noodles for chicken soup. He could see her, face hollowed by hunger, etched by worry, telling him how vital it was to look after this wallet. It contained his papers. In this world now, that meant it held his very soul, his permission to exist. She smiled and kissed him. "You behave now, Kurtl," she said. "Be a good child when you get there—no tricks, be obedient so they will let you stay." She produced a gift for him, a newly bought harmonica, all gleaming and sweet, and he clutched it to him . . .

. . . and then she was gone. Blinking out in his memory like a light switched off.

Kurt looked around at the people on the train, at the unfamiliar countryside flowing by under a February frost. He knew that this was the train from Berlin, where he had collected his final papers from the German Jewish Children's Aid and the required travel money—fifty crisp green American dollars tucked safe in his luggage—and he knew likewise that he had got to Berlin from Vienna on another train . . . but the memory of it was fading. In time, to his lifelong regret, he would be utterly unable to remember saying goodbye to his mother, or to Herta.

The old life, the familiar life, the beloved, was behind him, inexorably receding into a different dimension. Or perhaps it was the other way round—Vienna was real and of the present, and it was he who had been pushed into this unreal existence.

Most of the other people on the train were refugees, and to Kurt most seemed elderly. There were a few families with young children. German, Austrian, Hungarian Jews, a few Poles. Mothers murmured to their little ones while their husbands read or talked or dozed, old men with hats low on their brows stooped in their sleep, snoring and sighing into their beards, and children stared wide-eyed or drowsed against their parents.

Every few stops they all had to change trains, herded by police or soldiers onto whatever trains were available. Sometimes Kurt found himself in luxurious first-class compartments, sometimes second class, but more often the aching wooden slats of third. Kurt preferred the benches, because at least he got to sit properly; the seats in first had armrests, and the children had to perch on them, squeezed between the adults. On a few occasions Kurt got so desperate for comfort that he clambered up onto the luggage rack and stretched out on the valises.

There were only two other unaccompanied children on the train, a boy and a girl. Kurt gradually got to know them. One was a fellow Viennese named Karl Kohn, aged fourteen and from the same part of Leopoldstadt as Kurt. He wore glasses and seemed rather sickly, and was a little small for an adolescent. The girl could not have been more different; Irmgard Salomon was from a middle-class family in Stuttgart; despite being only eleven, she was taller than either of them by a clear two inches. Drawn together by their isolation, the three formed a bond as the train carried them farther and farther from their homes.

אמא

The apartment had become a hollowed shell. Where there had been family, now there were just two women: one aging, one just blooming. Tini was forty-seven years old—an age when she should have been looking forward to a future filled with grandchildren. And Herta, two months away from her nineteenth birthday, should have been settled in her occupation and considering which of her admirers she might marry. They should not have been sitting here alone in this desolate apartment, their few possessions robbed from them and their dear ones—husband, sons, daughter, father, brothers, sister—stolen or fled.

Vienna was a place of forbidden zones, and the apartment, which they were fortunate to have kept at all, was a prison.

Saying goodbye to Kurt had been a pain beyond pain. He was so small, so slight, such a sliver of humanity to be sent out into the world. Tini had not been able to accompany him to the train—only people with travel permits were allowed on the platforms—and she and Herta had had to say their farewells outside, watching from a distance as the crowd of refugees swept him away.[2]

Flesh of her flesh, blood of her blood, soul of her soul, gone from her. Kurt was her hope; he would have a new beginning in an altogether new world. Perhaps he would return one day, and she would see a new person in his place, shaped by a life that was wholly strange to her.

בן

Kurt lay on his back and gazed up at the stars. He had never in his life seen such a sky—deeper, darker, more brilliant than any other on earth: a vault unadulterated by man-made light. The ship, rolling steadily beneath him, was in blackout, alone on the vast disc of starlit black ocean.

He felt like the last survivor of a great exodus. After the train arrived in Lisbon, he and Karl and Irmgard had been kept waiting for weeks. There were supposed to be dozens of other children joining them for the voyage to America, but when the time came to sail, it was apparent that the others weren't going to make it. They were presumably trapped in the bureaucratic tangle of emigration. Kurt, Karl, and Irmgard were taken to the dock, where their ship waited, tall as an office block, moored alongside the quay with great ropes and gangways. SS *Siboney* wasn't the largest passenger liner afloat, but she had a certain elegance: two slender funnels and upper decks lined with arcaded promenades. Along the hull were identification markings to protect her from German U-boats: AMERICAN—EXPORT—LINES in giant white letters, flanked by the Stars and Stripes.

The majority of people aboard appeared to be refugees—many of them familiar faces from the train journey—with a few returning tourists and commercial travelers among them. Kurt and Karl went in search of their cabin, eventually finding it in the depths of the ship, where it was unpleasantly stuffy and the engines throbbed loudly. Returning to the fresh air, they watched as *Siboney* pulled away from the dock and, with engines thrashing the water to foam, turned her bow westward.

Kurt stood at the rail for three hours, looking out across the expanse of the ocean. Lisbon shrank to a smear, then Portugal to a sliver, then all of Europe dwindled and sank beneath the horizon. Out of sight, beyond the northern sea, convoy after convoy of merchant ships dragged slowly eastward toward Britain with Royal Navy escorts circling like nervous herds-men; in the east, U-boats slid out from their pens and cruised the vast ocean with torpedoes couched in their tubes. All *Siboney* had for protection was her painted markings.

Despite his tiredness, Kurt slept badly that first night in the noisy, overheated cabin, and the next day was marred by sea-sickness. All he could keep down was fruit. Reluctant to spend another night in their bunks, Kurt and Karl took their blankets and sneaked up on deck. There was nobody to stop them; Nurse Sneble, a compact middle-aged woman from New York, was supposed to look after the children, but she was occupied with the elderly passengers.

It was chilly in the night air, but wrapped up and reclining in deck chairs the two boys were warm enough. They luxuriated in the quiet and the fresh air. Kurt watched the stars overhead, wondering at this new situation and the place he was going to. He knew a tiny smattering of English from school; he could say *hello*, *yes* and *no*, and *OK*, but that was about all. His class had learned the rhyme "Pat-a-cake, pat-a-cake, baker's man" by rote, but in Kurt's mind the words had little meaning. To his ears, the Americans on board just spoke gibberish.

Somewhere back there, beyond where the eastern starfield met the black line of the ocean, were his home and family. The bright new harmonica, his last physical link with his mother, was gone. While he and the other kids were waiting to change trains somewhere in France, some German soldiers had chat-ted and played with them. Kurt had shown them the harmon-ica, and they took it and wouldn't give it back. Maybe they figured a Jew shouldn't have such nice things.

בן

A cloud lay over Europe, roiling and flickering with lightning. Somewhere in the mid-Atlantic, *Siboney* steamed out from under it and into a bright American dawn.

Kurt and Karl, asleep on their deck chairs, were woken by a dash of cold spray—not sea spray but a splash from the mop of a sailor swabbing the deck. Gathering their blankets, they retreated indoors.

Somehow Nurse Sneble found out about their night al fresco. They were reprimanded, and ordered to sleep in their cabin from then on. They continued to have the run of the ship all day long, exploring, playing games, making friends with the sailors, distracted for a while from what they had left behind and the uncertainty of where they were going.

After calling at Bermuda the ship turned northwest, leaving the warm tropics behind. Kurt sensed a changed atmosphere aboard; people were preparing for the most momentous arrival of their lives. Around noon on Thursday, March 27, 1941, with every man, woman, and child lining the rails, *Siboney* passed between Staten Island and Long Island.

Kurt pressed between the others to watch the gray waters and distant shores slip by. Off the port bow, the glittering outline of the Statue of Liberty grew from a little spike until she towered above the ship, pale green and magnificent. The ship steered into the Hudson past the skyscraper skyline of Manhattan. Children and adults chattered and pointed, wreathed in smiles. Many had been given little American flags and held them up, fluttering in the wind, tiny, fragile offerings of hope.

בן

Kurt's senses were nearly drowned in the immensity of New York. Canary-yellow taxicabs with flared black wings stuttered at the curbs and barked their way angrily into the streaming, screaming traffic, disputing the Forty-Second Street intersection with bell-ringing trams. Broadway and Times Square were like the innards of a racing engine with the throttle wide open. Kurt clutched the hand of the lady from the Aid Society like a lifebelt as they waded through the sidewalk crush of skirts and overcoats, swinging umbrellas and canes, flapping newspapers and flying cigarette ash.

This was nothing like Vienna. New York was all modernity from foundations to sky, a town built out of automobiles and glass and concrete and people and people and people and still more people who themselves seemed more of the modern world than any in Europe. Kurt and his friends were aliens in every way.

After *Siboney* had docked at the pier, and following a medical inspection,[3] the children had disembarked and been met by a lady from the Hebrew Immigrant Aid Society, which partnered with the German Jewish Children's Aid in helping refugees. Only Kurt had definite arrangements in place. Karl and Irmgard had no friends or relatives here; the charity had arranged places for Irmgard in New York and for Karl in distant Chicago. After a night in a hotel the time came for them to part. Kurt never saw either of his friends again.[4]

דוד

The strange place names ticked by, meaningless to Kurt's Austrian eyes. Each spoke of a previous wave of religious immigrants yearning for their hometowns: Greenwich, Stamford, Stratford,

Old Lyme, New London, Warwick. The railroad traced the coast through Connecticut to Providence, Rhode Island. There the main line ended.

When Kurt disembarked, accompanied by the suitcase that had traveled with him all the way from Im Werd, they were met by a woman about his mother's age, but more expensively dressed. To his surprise, she greeted him in German, introducing herself as Mrs. Maurer, his mother's old friend. Waiting with her on the platform was a middle-aged man accompanied by a woman, both regarding him with reserved benevolence. In respectful tones, Mrs. Maurer introduced the gentleman as Judge Samuel Barnet, Kurt's sponsor.

Judge Barnet was around fifty years old and rather short and stocky, with gray, receding hair, a large, fleshy nose, bushy eyebrows, and deceptively sleepy-looking eyes.[5] He had a rather grave demeanor, even a little frosty. The lady with him, who wasn't much taller than Kurt himself, was the judge's sister, Kate: neat and solidly built like her brother. Mrs. Maurer explained that Kurt wouldn't be staying with her; instead she had arranged accommodation with Judge Barnet himself.

From Providence they drove into Massachusetts, across a seemingly endless succession of rivers, bays, and inlets. Eventually they reached their destination: New Bedford, a large town on an estuary. This southeast corner of the state was a dense little patch of immigrant England whose traces were visible on almost every road sign for miles around, from here to Boston by way of Rochester, Taunton, Norfolk, and Braintree. All Kurt knew was that New Bedford was even less like Vienna than New York had been—a town of river ferries and small, genteel public buildings, cotton mills and long avenues of suburban homes of gray shingle and white clapboard, where automobiles hummed, children played, and sober citizens went about their business with decorum.

As both a pillar and keystone of the town—especially its Jewish community—Samuel Barnet might well be expected to be an intimidating presence, with an imposing mansion on the edge of town; instead, the car turned in at the driveway of a regular middle-class suburban house standing shoulder to shoulder with others almost but not quite identical to it.

Kurt's reception was warm but reserved. Communication was near impossible once Mrs. Maurer had left. "Pat-a-cake, baker's man" would be of no use at all in this situation. Fortunately, the judge wasn't alone in welcoming the new arrival. Samuel Barnet had been a widower for more than twenty years; with him lived his three middle-aged sisters, all resolutely unmarried. Kate, Esther, and Sarah appointed themselves Kurt's aunts, welcoming the bewildered boy and showing him his room. He'd never had a room of his own before.

Next morning he woke to find a strange presence at his bedside. A tiny boy aged about three, dressed in a little camel-hair coat, was gazing at him in wonder. The apparition opened its mouth to speak—and out poured a stream of incomprehensible English gibberish. The child seemed to want or expect something, but Kurt had no idea what. The boy's face fell in disappointment, and he burst into tears. He turned to an adult standing behind him and wailed, "Kurt won't talk to me!"

The little boy, Kurt learned, was David, the son of Judge Barnet's younger brother, Philip, who lived next door. Together they made up one large extended household. Over the next few weeks, Kurt was quickly and smoothly assimilated. Uncle Sam—as Kurt learned to call Judge Barnet—belied his somber appearance and proved as warmly welcoming as any guest could wish. Kurt would never be allowed to feel out of place. Years later he would learn how fortunate he had been; not all refugee children landed on their feet. Many found themselves in unfriendly households or suffered anti-Semitic or anti-German abuse in the neighborhood, or both. As Kurt got to know New Bedford, he would discover that the Barnets were leading lights in a large Jewish community, within which he was welcomed.

The Barnet family were Conservative Jews. All Kurt had known were his family's lightweight religious observances, in which synagogue and Torah played little roles, and the strictly Orthodox who were common around Leopoldstadt. Conservatives—who were not necessarily politically conservative—were somewhere in between; they believed in preserving ancient Jewish traditions, rituals, and laws, but departed from the Orthodox in recognizing that human hands had written the Torah and that Judaic law had evolved to meet human needs.

Spring was coming to New Bedford, and the trees lining the street turned green. If you squinted along it, you could almost imagine that you were in the Hauptallee in the Prater, and that none of this had happened—the Nazis coming, the family sundered. Kurt could already sense—if it hadn't been for the lack of his mother and father, and of Fritz, Herta, and Edith, and the vast distance that lay behind him—that he had found something that felt like a home.

8. Unworthy of Life

אח

Nobody ever knew the cause of Philipp Hamber's murder, but everyone heard about the circumstances. The SS required no reasons for their brutalities: a bad mood, a hangover, a prisoner looking askance at a guard, or just a sadistic impulse. When SS-Sergeant Abraham knocked Philipp Hamber to the ground and killed him, what the witnesses remembered was the atrocity itself, and the terrible repercussions for them.[1]

"Again there is unrest in the camp," Gustav wrote. He rarely took his diary from its hiding place these days. His last entry had been in January 1941, when they were shoveling snow. Now it was spring. In the intervening months the prisoners had been growing less submissive to SS violence.

At the end of February, a transport of several hundred Dutch Jews had arrived. There had been violent clashes in the Netherlands between homegrown Dutch Nazis and the Jewish population, and in Amsterdam the Nazis suffered a severe beating at the hands of young Jews. The SS rounded up four hundred as hostages, a move that triggered a wave of strikes, paralyzing the docks and triggering open warfare between the strikers and the SS. At the end of the month, 389 of the Jewish hostages were transported to Buchenwald.[2] Some were quartered in block 17, and Fritz spent a lot of time with them. He and his friends tried to teach the Dutchmen the ways of the camp, but it did them little good. They were strong, spirited men who weren't easily cowed, and the SS treated them with an unprecedented level of brutality. All were put to work as stone carriers in the quarry, and in the first couple of months around fifty were murdered. Deciding that the Dutchmen couldn't be broken quickly enough, the SS shipped the survivors off to the notoriously brutal Mauthausen. None ever returned.

The Dutchmen left behind them a budding spirit of defiance inspired by their resilience. When Philipp Hamber was murdered, the prisoners' mood began to simmer dangerously.

Like Gustav, Philipp was Viennese and worked in the haulage column, in a different team under a kapo called Schwarz. His brother, Eduard, was in the same team. Philipp and Eduard had been movie producers before the Anschluss. Despite being unused to physical labor, they had survived three years in Buchenwald. On this particular spring day, their team had made a delivery to a building site. SS-Sergeant Abraham, one of the cruelest, most feared Blockführers in Buchenwald, happened to be there.[3] Something—a misdirected glance from Philipp, a mistake, perhaps a dropped sack of cement, or just something about the way he looked or moved—drew the SS man's attention.

In a rage, Sergeant Abraham shoved Philipp to the ground and kicked him. Then he seized the helpless man's collar and dragged him through the churned mud of the building site, heaving him into a foundation trench full to the brim with rainwater. As Philipp floundered and choked, Abraham planted a boot on the back of his head, forcing him beneath the surface. Eduard, along with the other prisoners, watched in silent horror as his brother struggled. Philipp's thrashing gradually subsided and his body went limp.

Buchenwald was accustomed to murder as a constant part of its everyday life, the prisoners learning to live with it and to avoid it as best they could. But now they were becoming resentful. News of the killing of Philipp Hamber spread like a flame.

Gustav brought his long-neglected diary out of hiding and set down how Philipp had been "drowned like a cat" and that the prisoners were not taking it quietly. Much of the unease and anger came from Eduard.[4] He wanted justice for his brother.

His cause was helped by the fact that the murder, having occurred on a construction site in the SS complex, had been witnessed by a civilian visitor; therefore Commandant Koch had no option but to enter the death in the camp log and hold an inquiry. Simultaneously, Eduard lodged an official complaint. He was aware of the danger he was putting himself in. "I know that I must die for my testimony," he told a fellow prisoner, "but maybe these criminals will restrain themselves a little in the future if they have to fear an accusation. Then I will not have died in vain."[5]

He had underestimated the SS. At the next roll call, all Philipp's comrades from kapo Schwarz's haulage detail, including Eduard, were called to the gatehouse. Their names were taken, and they were asked what they had witnessed. Terrified, they all denied having seen anything. Only Eduard persisted in his accusation. While the others were sent back to their blocks, Eduard was interrogated again by Commandant Koch and the camp doctor. Koch assured him, "We want to know the whole truth. I give you my word of honor that nothing will happen to you."[6] Eduard repeated his account of how Abraham had attacked his brother and deliberately, brutally drowned him.

They let him go back to his block, but late that night he was called out again and taken to the Bunker—the cell block that occupied one wing of the gatehouse. The Bunker had an evil reputation; tortures and murders were perpetrated in there, and no Jew who entered it ever came out alive. Its principal jailer and torturer was SS-Sergeant Martin Sommer, whose boyish looks belied years of experience in concentration camps. Everyone knew Sommer well from his regular performances wielding the whip when victims were taken to the *Bock*.

After four days in the Bunker, Eduard Hamber's corpse was brought out.

It was claimed that he had committed suicide,[7] but it was common knowledge that Sommer had tortured him to death.

This wasn't enough to satisfy the SS. At intervals over the following weeks, three or four of the witnesses from the Schwarz detail would be named at roll call and brought to the Bunker. There they were interrogated by Deputy Commandant Rödl (the music lover) and the new camp physician, SS-Doctor Hanns Eisele. The prisoners were told that they had nothing to fear if they told the truth. Knowing perfectly well that this was a lie, they continued to deny that they had seen anything. Their silence did not save them; they were murdered to the last man.

Gustav described the successive disappearances in his diary; the men were marched to the Bunker "and taken care of by Sergeant Sommer: even Lulu, a foreman* from Berlin, and (so kapo Schwarz believes) Kluger and Trommelschläger from Vienna are among the victims. Thus our rebellion shrivels up."[8]

Eduard Hamber had based his heroic sacrifice on the premise that the SS could be brought to account for their crimes, or at least be made to fear that they might be. All he had proved was that they were immune and their power was limitless.

אמא

Tini sat at the table where her family had once eaten together. "My beloved Kurtl," she wrote. "I am extremely happy that you are doing fine and are well. I am really curious to hear about your summer vacation. Actually, I almost envy you; one cannot go anywhere anymore here . . . I would be so glad if I could be with you now. Here, we cannot enjoy ourselves anymore . . .'"[9]

---

* Foreman was a semi-unofficial designation beneath kapo in rank.

Restrictions on Jews had been tightened still further in May with a declaration reinforcing and extending existing laws: Jews were forbidden to visit all theaters, concerts, museums, libraries, sporting events, and restaurants; they were barred from entering shops or buying goods outside specified times. Whereas they had been forbidden to sit on designated public benches, they were now barred entirely from public parks. The declaration also introduced some new rules: Jews were not allowed to leave Vienna without special permission and were banned from making enquiries to the government. The spreading of rumors about resettlement and emigration was strictly prohibited.[10]

Tini still hadn't given up her efforts to get Herta and Fritz to America. But it was harder than ever. Shortly after Kurt's departure, Portugal had suspended transmigration due to a bottleneck at Lisbon, and in June, President Roosevelt stopped the transfer of funds from the United States to European countries, hamstringing the refugee aid agencies.[11] In the first half of 1941, only 429 Viennese Jews had managed to emigrate to the United States, leaving behind 44,000 desperate to escape.[12] Then, in July, US immigration regulations invalidated all existing affidavits.[13]

All Tini's plans were crushed. But still she went on trying. It wore her down; some days the depression weighed so heavily on her that she couldn't drag herself out of bed. Just recently, news had come to several neighboring families that their menfolk had died in Buchenwald, all persecuted to the point where they committed suicide by running through the sentry line. Every day Tini expected to hear similar news about Gustav and Fritz. It tormented her to know the kind of grueling labor her husband was made to do—"He is not a young man anymore," she wrote. "How can he bear that?"[14] Every time a letter from them was delayed, it sent her into a panic. So she persevered and fought on, refusing to give up hope of at least getting Herta to safety. With the tiny sums of money she could scrape together, the necessary fees, taxes, and bribes were virtually impossible. She'd had a brief stint working in a grocery store, but had been fired because as a Jew she was not a citizen.

"Life is getting sadder by the day," she wrote to Kurt. "But you are our sunshine and our child of fortune, so please do write often and in detail . . . Millions of kisses from your sister Herta, who is always thinking of you."[15]

בן דוד

Judge Barnet hadn't tarried in putting Kurt to school, despite his speaking no English. He picked up the language quickly, thanks in large part to coaching from Ruthie, the Barnets' niece, who came to live with them that summer.

Ruthie had graduated from college and taken a job as a teacher at Fairhaven, across the estuary from New Bedford. Each day when Kurt came home from school Ruthie tutored him in English. She was a fine teacher, kind and good-natured, and Kurt grew to adore her; in time she would become a sister to him, in place of Edith and Herta. Cousin David next door would become a little brother, their relationship echoing Kurt's bond with Fritz.

In those first months, Kurt was photographed for the local newspaper, interviewed on the radio, and when he graduated fourth grade in June, the teacher placed him front and center in the class photograph. That first summer, when he was still finding his feet, he was sent to Camp Avoda, a summer camp founded by Sam and Phil Barnet that took Jewish boys from deprived urban environments and gave them a grounding in traditional values.

The camp was set among the trees on the shore of Tispaquin Pond, between New Bedford and Boston, a group of utilitarian dorm huts surrounding a baseball field. Kurt had the time of his life, playing sports and swimming in the warm, shallow waters of the lake; in Vienna he had floundered in the Danube Canal with a rope tied round his waist and a friend on the bank hold-ing the other end; here he learned to swim properly. Had Fritz been able to see Camp Avoda, he might have been reminded of the paradise described in Makarenko's *Road to Life*.

Normally Kurt didn't like to write letters, but now he wrote profusely to his mother, telling her all about this wonderful new world he had found.

Tini devoured every detail of his news, heartened to know that two of her children were now safe. (She assumed Edith was all right, despite having been out of contact for nearly two years now.) But she couldn't shed her anxiety that something would go wrong, that somehow Kurt's idyll would be destroyed. "Please be obedient," she pleaded, "be a joy for your uncle, so that the counselors have good things to say about you . . . Darling, please be well behaved." A photograph he sent her with the other Bar-net children filled her with pleasure: "You look so nice . . . so handsome and radiant. I almost didn't recognize you."[16]

Kurt was losing his old life in the brightness of the new.

אבא

Summer returned to the Ettersberg. "Fritzl and I are now receiving money regularly from home," Gustav wrote. It was little, but it helped make life bearable. Tini also sent occasional packages of clothing—shirts, underpants, a sweater—which were invaluable. Whenever a packet arrived, Gustav or Fritz would be called to the office to collect and sign for it, the con-tents itemized on their record cards.[17]

Gustav's love for his son had grown to fill his whole heart during their time in Buchenwald. So had his pride in the man Fritz was becoming—this June he would turn eighteen. "The boy is my greatest joy," he wrote. "We strengthen each other. We are one, inseparable."[18]

On Sunday, June 22, the camp loudspeakers announced momentous news. That morning, the Führer had launched an invasion of the Soviet Union. It was the biggest military action in history, with three million troops on a front spanning the whole of Russia intended to engulf it in one huge wave.

"Every day the roar of the radio," Gustav wrote. The camp loudspeakers, always an intermittent source of unwelcome noise—blaring out Nazi propaganda, German martial music, terrifying commands, and morale-grinding announcements—now played an almost constant stream of Berlin radio, crowing with triumphal news from the eastern front. The glorious crushing of Bolshevik defenses by the might of German arms, the encirclement of Russian divisions, the seizing of city after city, the crossing of rivers, the victory of some Waffen-SS corps or Wehrmacht general, the surrender of hundreds of thousands of Soviet soldiers. Germany was devouring the lethargic Russian bear like a wolf disemboweling a sheep.

For Jews under Nazi rule—especially those in the Polish ghettos—the invasion of the Soviet Union gave a glimmer of hope; Russia might win, after all, and liberate them from this miserable existence. But to the political prisoners in the concentration camps, most of whom were communists, the news of Soviet defeats was depressing. "The politicals hang their heads," Gustav noted.

Unrest was stirring among the prisoners again. There were disturbances in the labor details, incidents of disobedience, minor acts of resistance. The SS dealt with it in their usual way. "Each day the shot and slain are brought into the camp," wrote Gustav. Each day, more work for the crematorium, more smoke from the chimney.

In July, a new horror came to Buchenwald, a foreshadowing of the future. It was supposed to be veiled in secrecy, but the veil was thin.

The previous September, an American journalist in Germany had reported a "weird story" told to him by an anonymous source: "The Gestapo is now systematically bumping off the mentally deficient people of the Reich. The Nazis call them 'mercy deaths.'"[19] The program, code-named T4, involved specialized asylum facilities equipped with gas chambers, together with mobile gas vans which traveled from hospital to hospital, collecting those deemed by the regime "unworthy of life." Negative public attention, particularly from the Church, had led to the T4 program being suspended. In its place, the Nazis began applying it to concentration camp inmates. This new program, code-named Action 14f13, was to focus particularly on disabled Jewish prisoners.[20] In Buchenwald, Commandant Koch received a secret order from Himmler; all "imbecile and crippled" inmates, especially Jews, were to be exterminated.[21]

The first the inmates of Buchenwald knew of Action 14f13 was when a small team of doctors arrived in the camp to inspect the prisoners. "We got orders to present ourselves at the infirmary," Gustav wrote. "I smell a rat; I'm fit for work."[22]

One hundred and eighty-seven prisoners were selected, variously classed as mentally handicapped, blind, deaf-mute, or disabled, including some injured by accidents or abuse. They were told that they would be going to a special recuperation camp, where they would be properly looked after and allotted easy work in textile factories. The prisoners were suspicious, but many—especially those most in need of care—chose to believe the hopeful lies. Transports came and collected the 187 men. "One morning, their effects came back," wrote Gustav. The grim delivery included clothing, prosthetic limbs, and spectacles. "Now we know what game is being played: all of them gassed." They were the first of six transports of prisoners murdered under Action 14f13.

At the same time, Commandant Koch began an ancillary

program: the elimination of prisoners carrying tuberculosis. SS-Doctor Hanns Eisele was in charge. A virulent anti-Semite, Eisele was known to the prisoners as the *Spritzendoktor*—Injection Doctor—because of his keenness to dish out lethal injections to sick or troublesome Jews. He was also known as White Death,[23] using prisoners for vivisection for his own personal edification, administering experimental injections and unnecessary surgery—even amputations—and then murdering the victims.[24] He would be remembered as perhaps the most evil doctor ever to practice at Buchenwald.

The scheme began when two large transports arrived carrying prisoners from Dachau. Five hundred were diagnosed with tuberculosis—on the basis of general appearance rather than a proper medical examination—and sent to the infirmary. There they were immediately killed by Dr. Eisele with injections of the sedative hexobarbital.[25]

Within a few months, the character of Buchenwald had altered irrevocably. From now on, anything which weakened a man—any injury, sickness, or disability—was as good as a death sentence. Such things had always carried a severe risk, but now it became a stone certainty that being rated unfit for work or "unworthy of life" automatically put a man's name on an extermination list.

And then the first Soviet prisoners of war arrived, and yet another new door opened into another new department of hell.

In the Nazi mind, Jews and Bolsheviks were one and the same—Jews, they claimed, had created and spread communism, and now ran it alongside the global capitalist conspiracy that they were also, contradictorily, alleged to be running.[26] This mythology had inspired the invasion of the USSR and a campaign of murder, with death squads following the army and slaughtering Jews in their tens of thousands. Captured Red Army soldiers, hundreds of thousands of whom had been rounded up in the first weeks of the invasion, were treated as subhuman—if not Jews, then the thralls of Jews: degenerate and dangerous. Political commissars, fanatical communists, intellectuals, and Jews were singled out for immediate disposal. The task couldn't be accomplished in the POW camps because of the risk of spreading panic among the bulk of the prisoners; thus the SS decided to use the concentration camps. The program was code-named Action 14f14.[27]

בֵּן

On a day in September, Fritz stood at roll call in the column with the other block 17 men. His papa stood with the men of his barrack in another part of the square.[28] It was like every one of the hundreds of other roll calls they had stood through. The tedious progression of numbers and answers; the announcements; the round of routine punishments . . . and then, something entirely unprecedented.

That day, the first transport of Soviet prisoners of war had arrived in Buchenwald. They were a small batch: just fifteen lost, frightened men in tattered Red Army uniforms. Fritz watched curiously as Sergeant Abraham (Philipp Hamber's killer) and four other guards surrounded the Russians and marched them off the square. Several thousand pairs of eyes followed them as they went. At the same time, the camp orchestra sat tuning up. On an order from the podium, they began to play the "Buchenwald Song."

The routine of roll-call singing was so ingrained that Fritz and his comrades opened their mouths and gave voice without thinking:

> When the day awakens, ere the sun smiles,
> The gangs march out to the day's toils . . .

Straining his eyeballs, Fritz watched the Russians being force-marched past the crematorium toward the section of the camp occupied by a small factory—the Deutsche Ausrüstungs-werke (DAW), whose prisoner workforce manufactured military equipment for the German army—beyond which was an SS shooting range. The POWs and their guards passed out of sight.

> And the forest is black and the heavens red,
> In our packs we carry a scrap of bread,
> And in our hearts, in our hearts, just sorrow.

Thousands of voices roared over the camp, almost but not entirely drowning out the volleys of shots from beyond the factory.

The Russian soldiers were never seen again. A couple of days later, another thirty-six Soviet POWs were brought to the camp, and again the prisoners had to sing to drown out the gunshots.

"They say they were commissars," Gustav wrote, "but we know everything . . . How we feel is not to be described—now shock is piled upon shock."

This method of execution proved too inefficient to handle the large numbers of Russians the SS wished to dispose of. Thus, while small groups were being murdered on the shooting range, a new facility was being prepared. In the woods near the road to the quarry, the SS had a disused stable building, in which a team of carpenters from the construction detail were hard at work. The facility was code-named Commando 99, and its purpose, though secret, would soon become apparent.[29] At the same time, three barrack blocks in the corner of the main camp were fenced off, forming a special enclosure for Soviet POWs, who began to arrive in their thousands.[30]

Each day, Russians selected for liquidation were taken in groups to Commando 99, where they were told they would undergo a medical inspection. They were led, one at a time, through a series of rooms filled with medical paraphernalia and staffed by men in white coats. The prisoner's teeth were examined, his heart and lungs listened to, his eyesight tested. Finally he was led into a room with a measuring scale marked on the wall. Obscured by the scale was a narrow slit at neck height, behind which was a concealed room in which stood an SS man armed with a pistol. While the prisoner was being measured, the attendant tapped on the partition, and the hidden guard shot the prisoner in the back of the neck.[31] Throughout the building, loud music drowned out the sounds of the shots, and while the next victim was being brought through, the previous prisoner's blood was hosed off the floor.

Fritz and Gustav and all their fellow prisoners knew perfectly well the nature of the "adjustments" (as the SS officially called the executions) being carried out in the old stable.[32] The carpenters who had converted the building were Fritz's workmates. Truckloads of Russians arrived daily and disappeared; and everyone saw the closed van driving up the hill from Commando 99, dribbling trails of blood along the road and across the square to the crematorium. After a while, the van was fitted with a metal-lined container to prevent leakage. The crematorium couldn't cope with the numbers, and mobile ovens had to be brought up from Weimar; they were parked on the edge of the roll-call square, incinerating the bodies right in front of the other prisoners.[33]

"Meanwhile the shootings continue," Gustav recorded.

אחים

Surely one must finally lose the ability to be appalled? It must get worn down like a stone with the passage of use, blunted like a tool, numbed like a limb. One's moral sense must scar and harden under an unending series of lacerations and bruises.

For some, perhaps that was so; for others, the opposite was true. Even some of the SS could only withstand so much. The camp guards all had to take turns handling the victims in Commando 99 and wielding the pistol, and found that continuous, orchestrated butchery was not the same as the sporadic murders they were accustomed to. Many reveled in it; they saw themselves as soldiers, and these killings were their contribution to the war against Bolshevik Jewry, but others were broken by it, and tried to avoid duty in Commando 99; some fainted when faced with the carnage or suffered mental breakdowns; a few worried that if word got out—as it inevitably would—it could lead to retaliation against captured German troops by the NKVD, the Soviet Gestapo.[34]

For the Buchenwald prisoners, all of whom were witnesses to Action 14f14 and some of whom were forced participants in the cleaning up, the effect was corrosive and traumatic. And it was far from being the end.

At the end of 1941, prisoners began to be subjected to lethal medical experiments designed to develop vaccines for German troops.

Everyone knew that something was afoot when they fenced off block 46—one of the two-story stone-built barracks near the vegetable gardens. After roll call one winter's day, the adjutant produced a list and stood surveying the massed ranks of prisoners before beginning to call out numbers. The heart of every man there beat a little faster; whenever the SS compiled a list, it was never for anything good. Each selected man turned pale as his number was called.

It was doubly unnerving that SS-Doctor Erwin Ding* was on hand. A trim, nervous-looking little man who had served with the Waffen-SS, Ding was known for his incompetence.[35]

* Later known as Schuler or Ding-Schuler.

The same was true of his deputy, SS-Captain Waldemar Hoven; a remarkably handsome fellow, Hoven had worked as a movie extra in Hollywood; medically unqualified, he was even more incompetent than Ding. But he was very handy at delivering lethal injections of phenol.[36]

The prisoners whose numbers had been called—a mixture of Jews, Roma, political prisoners, and green-triangle men—were marched to block 46 and disappeared inside.

What happened to them in there only became known when the survivors were let back out. Ding and Hoven injected the prisoners with typhus serums; they immediately fell ill with bloating, headaches, bleeding rashes, hearing loss, nosebleeds, muscle pain, paralysis, abdominal pain, vomiting. Many died, and the survivors were left in a pitiable state.[37]

At periodic intervals, more batches of prisoners were sent to block 46 to be ruined and killed in the name of research. Several old friends of Gustav's from Vienna were among the prisoners selected for torment. However, they were saved when the SS high command deemed it improper for Jewish blood to be used in the development of a vaccine that was to be injected into the veins of German soldiers. The Jewish subjects were ejected from the program and returned to the normal hell of the camp.[38]

אם וכת

Tini and Herta sat at the kitchen table, plying their needles and thread. Mending had always been a part of Tini's married life; with little income and four children, there had always been stitching and darning to do. Now, month by month, Herta's and her own clothes got shabbier, and their needles worked overtime to keep them in one piece.

Today, however, their sewing was not mending. On September 1, 1941, it had been announced by the Ministry of the

Interior in Berlin that as of the nineteenth of the month, all Jews living in Germany and Austria must wear a yellow Star of David on their clothing—the *Judenstern*.

The Nazis had already revived this medieval practice in Poland and other occupied territories. Now it had been decided that *all* Jews, including those at home, must be deprived of their ability to be camouflaged within society.[39]

Along with their neighbors and relatives, Tini and Herta had had to go along to the local IKG collection point to get their stars. They were factory-made, printed on rolls of fabric, with the word *Jude* in black lettering styled to resemble Hebrew.[40] Each person was allotted up to four. The final insult was that they had to pay for them: ten pfennigs each. The IKG bought them in huge rolls from the government for five pfennigs a star and used the profit to cover administrative costs.[41]

Even now, Tini hadn't given up the fight to get Herta away from this nightmare. There were girls her age and even younger being sent to concentration camps now. In desperation, Tini had written to Judge Barnet in America, begging him to help. Despite his offer of sponsorship, the usual obstructions had blocked Herta's visa. "I am devastated that she has to stay here. I was informed by an unofficial source that relatives in the US can petition Washington to obtain a visa. May I ask you to do something for Herta? I do not want to have to reproach myself like in Fritz's case."[42] Sam Barnet had acted right away, filing the necessary papers and putting up $450 to cover all Herta's expenses.[43] But the bureaucratic maze had been too complex and the barriers impossible to surmount. Herta's visa had not been approved.

Their needles plied in and out, through the cheap yellow calico of the stars and the worn wool of their coats. Tini glanced across at her daughter; she was fully a woman now—nineteen, going on twenty, about the age Edith had been when she went away. Nineteen and pretty as a picture. Imagine how beautiful she could have been if there were nice clothes for her and not this life of deprivation and fear. And when Herta looked back at her mother, she saw lines etched by worry and cheeks sunken from hunger.

The appearance of the yellow stars in Vienna over the following weeks produced strong reactions among non-Jews. They had grown so used to the idea that Jews had largely disappeared from the country—vast numbers had emigrated, and the supposedly dangerous ones had been sent to the camps—that it was as if thousands had suddenly materialized in their midst, marked for all to see. Some people were ashamed of what the Nazis had done; they believed that it was right and proper to bar Jews from public life, but to stigmatize them in this highly visible way was somehow wrong. Shopkeepers who had been willing to sell discreetly to Jews now had the embarrassment of having their other customers know that they did so. Some braved it out; others began to shut their doors to wearers of the yellow star. For those Jews who had been sufficiently Aryan-looking to ignore some of the restrictions, that was now out of the question. Some members of the public, shocked to find so many Jews still about, began to demand that harsh action be taken.[44] It seemed that life could not possibly get any worse.

But of course it could; the bottom of the pit had not yet been reached, not by any means.

On October 23, the head of the Gestapo in Berlin relayed an order to all Reich security police. With immediate effect, all emigration of Jews was banned.[45] Their removal from the Reich would now be solely by compulsory resettlement to newly established ghettos in the eastern territories. The last vestiges of Tini's hopes for Herta were snuffed out with the stroke of a bureaucrat's pen.

In December, in the wake of Pearl Harbor, Germany declared war on the United States, and the final barrier fell.

9. A Thousand Kisses

אבא

Spring had come to Buchenwald again; Gustav's and Fritz's third. The forest was alive with greenery, the singing of black-birds counterpointing the harsh *scrawk* of the crows. Each morning, soon after the rising of the sun, would come the rasp of saws biting into tree trunks, the grunts of the slaves wielding them, and the snapped insults and orders of the kapos and guards. Then a yell, and a great beech or oak would come crashing down, the slaves setting about it, reducing it quickly to logs and a carpet of leaves.

Gustav, already tired, shoulders raw from carrying, went among them with his team, collecting logs to transport to the construction sites. He was doing well—a foreman now, in charge of his own twenty-six-man team. "My lads are true to me," he wrote, "we are a brotherhood, and stick tightly together." Friendship was precious, and often short-lived. In February several of Gustav's friends, "all strong fellows," had been sent away in another transport of "invalids," and next day there had been the usual returning crop of clothes, prosthetics, and spectacles. "Everyone thinks, *tomorrow morning it will be my turn.* Daily, hourly, death is before our eyes."

In February, the SS had murdered Rabbi Arnold Frank-furter, who had married Gustav and Tini in 1917, flogging and tormenting him until his aged body could take no more. In the wreck that remained, it was hard to recognize the portly, bearded rabbi of old Vienna. Before he died, Rabbi Frankfurter asked a friend to pass on a traditional Yiddish blessing to his wife and daughters: *"Zayt mir gezunt un shtark"*—"Be healthy and strong for me."[1] Gustav remembered his wedding day clearly, in the pretty little synagogue in the Rossauer Kaserne, the grand army barracks in Vienna: Gustav in dress uniform, the Silver Medal for Bravery gleaming on his breast; Tini in picture hat and dark coat, almost plump before decades of hardship and mothering sculpted her into handsome maturity.

Taking off his cap and running a hand over the bristles of his shaved scalp, Gustav looked up into the canopy of swaying leaves. With a feeling like a faint ghost of contentment, he replaced his cap and sighed. "In the forest it is wonderful," he had written in his diary. "If only we were free; but always we have the wire before our eyes."

Work these days was even more exhausting than ever; since January a new commandant had taken over: SS-Major Hermann Pister. "From now on a new wind blows in Buchenwald," he'd told the assembled prisoners,[2] and he meant it. An exercise regime had been introduced, in which prisoners were roused half an hour earlier than usual for roll call and made to do exercises half dressed.

Hitler's hatred of Jews was swelling beyond all control or constraint. The invasion of the Soviet Union had failed to achieve the decisive conquest he'd expected. A food crisis had taken hold in the Reich, and communist partisans were causing trouble everywhere from France to the Ukraine. In the fevered Nazi mind, it was all the fault of the Jews; having caused the war in the first place with their global conspiracies, they were now hobbling German progress.[3] In January 1942, the heads of the SS had agreed at last upon the Final Solution to the Jewish problem. Mass deportation, emigration, and incarceration had not worked. Something far more drastic and decisive was required. The exact nature of it was kept secret from the public, but it transformed the concentration camp system. Jews came under even closer, even more hostile attention than before. In Buchenwald, euthanasia of invalids, starvation, abuse, and murder had whittled down the Jewish prisoner population, until by March, there were only 836 left, among a total of over eight thousand other prisoners.[4] The only thing keeping Buchenwald's remaining Jews alive was their usefulness as workers, and that might not hold out for long under the pressure from the top to bring about a "Jew-free Reich."

Gustav's momentary idyll, gazing up at the swaying trees, ended abruptly. Under his direction, his team lifted and shouldered the logs. (They had no wagon for this job; the timber had to be transported by hand up the thickly wooded hillside.) Gustav took great care distributing the loads; he was conscious that some of the men were too worn out to survive another trek up the hill with a tree trunk gouging their shoulders. He instructed them quietly to tag on with the others; so long as they were discreet and *looked* like they were carrying, they should be OK. Gustav shouldered his own end of a log, and they set off.

Approaching the building site, in view of a construction kapo and SS overseer Sergeant Greuel, the men forced themselves to pick up the pace. The last few yards and the stacking of the logs were done at top speed. This was dangerous; men had been maimed and killed by hastily stacked trunks slipping and rolling on them.[5]

"What do you think you're doing, Jew-pigs?"

Sergeant Greuel's apoplectic face appeared in front of Gustav; he pointed his hefty cane. "Some of these beasts aren't carrying anything!"[6]

Gustav looked at his men; they hadn't been as careful as he'd asked. They could hardly be blamed; they were worn half to death. "I'm sorry, sir. Some of my lads are worn ou—"

Greuel's cane lashed him across the face, knocking him sideways. Gustav put up his hands to protect his head, but the cane whipped furiously back and forth, battering his fingers. He twisted, and the blows fell on his back. As Gustav dropped to the ground, Greuel turned his rage on the other men, striding among them, beating them until they bled. When the storm had spent itself, he turned back to Gustav, breathing heavily from his exertions. "You're a foreman, Jew," he said. "Drive your Jew animals harder. I'll make a report about this lapse."

The next day it happened again—Gustav and his men were beaten for supposedly not working hard enough. At roll call, Gustav was called to the gate and interrogated by the Rapportführer, the sergeant who oversaw roll calls and handled camp discipline. By SS standards he was a reasonable man, and satisfied by Gustav's answers, he tore up Greuel's report.

But Greuel wasn't deterred. He was a sadist. Some believed there was a sexual element in his cruelty; he was known to hold individuals back from work details and beat them alone in his room for his own pleasure.[7] Once he'd fixed on a victim, he wouldn't let go. On the third day, Gustav and his team were hauling stone from the quarry. Their wagon was loaded with two and a half tons of rocks, and even with twenty-six men at the ropes it was a killing strain to drag it step by step to the top of the hill. Greuel was watching, and submitted another report against Gustav for not driving his team fast enough. This time the Rapportführer passed the report on for further action.

At roll call Gustav was summoned to the gatehouse again. For his dereliction of duty he was sentenced to five Sundays on the punishment detail, without food. Like Fritz before him, he was put on *Scheissetragen*—shit-carrying. Each Sunday, while most prisoners were taking it easy, he carried buckets of feces from the latrines to the gardens, always at a running pace. He was fifty-one years old and, tough as he was, his body surely couldn't take this treatment for much longer. His friends slipped him morsels of food on punishment days, but he lost twenty-one pounds in the course of a month. He'd always been lean; now he was becoming skeletal.

Eventually his sentence was completed and he returned to work. He was relieved of his position as foreman on the haulage column, but his friends managed to get him less arduous work on the infirmary wagon, carrying food and supplies. However, he still had to work evening shifts on haulage. He began to recover from his ordeal. That he had survived Greuel's persecution at all was little short of miraculous. Without his own strength of spirit and the support of his friends, it would have destroyed him as it did so many others.

בן

Fritz had learned that even miracles couldn't last in a place like this. Every day the circle of probability was closing in on each man, his days shortening and the odds of his survival growing longer.

That spring Fritz had lost one of his dearest friends, Leo Moses, the man who had protected and tutored him in the art of survival, who had secured safer employment for both Fritz and his papa. A large transport of prisoners had been sent to a new camp being built in Alsace, called Natzweiler. Leo was sent with them. Fritz never saw him again.[8]

One evening in June, Fritz was sitting in his regular place at the table in block 17, listening to the conversation of the older men. They had finished dinner—a small ration of turnip soup and a piece of bread—and fell to talking. Fritz listened keenly, but was too in awe of them to participate. He would be nineteen in a couple of weeks, still a boy in years and, in comparison with these men, a child in intellectual development and understanding of the world. Keen to learn, he soaked up their political discussions, tales of show business, and grand schemes for the future of Europe.

Fritz's attention was caught by the appearance of a familiar figure in the doorway. He looked up and saw kapo Robert Siewert, beckoning. Leaving the table and stepping outside into the mild evening air, Fritz found Siewert looking grave. He spoke softly and quickly: "There's a letter from your mother in the mail office. The censor won't let you have it."

Siewert was part of the prisoner network, with contacts in all the administrative offices where trusted prisoners were employed—including the mail room. He had managed to learn the letter's contents. The summer warmth shrank away from Fritz as the words sank in. "Your mother and your sister Herta have been notified for resettlement. They've been arrested and are waiting for deportation to the east."[9]

In a panic, Fritz rushed down the street to his father's block, with Siewert hurrying behind. Some inmates were hanging about outside, and Fritz asked them to tell his papa that he wanted to see him urgently. (Prisoners were forbidden to enter blocks other than their own.) After a few moments, Gustav came out. "Tell him," said Fritz, and Siewert repeated his summary of Tini's letter.

Resettlement, deportation. They could only speculate what it meant. Rumors circulated all the time, and they had acquired an acute sensitivity to Nazi euphemisms. Fritz and Gustav had heard the whispers about SS massacres in the Ostland, the conquered region east of Poland.[10] One thing at least was certain—there would be no more letters, no more link with Tini and Herta once they were gone from Vienna and sent to Russia or who knew where.

אם וכת

Tini stood by the gas cooker in the kitchen. She recalled the day they took Fritz and she threatened to gas herself if Gustav didn't run away and hide. A lot of good that threat had done. And now they had come for her.

She turned off the main gas tap, as she was required to do. The detailed list of instructions issued by the authorities lay on the kitchen table, along with the key ring with which she had been provided, with the apartment key attached to it.

Herta stood by in her patched coat with the yellow star on the breast, her little suitcase by her side. They were permitted only one or two bags per person—total not to exceed one-hundred and ten pounds. They had packed clothes and bedding—as required in the resettlement instructions—along with plates, cups, and spoons (knives and forks were forbidden), and food to last for three days' travel. Those who had them were required to bring equipment and tools suitable for establishing or maintaining a settlement. Tini would be permitted to keep her wedding ring, but all other valuables had to be surrendered. She'd never possessed many treasures, and they were all gone now, anyway, stolen from her or sold; neither could she have conjured up more than a fraction of the three hundred marks in cash that the deportees were allowed to take to the Ostland.[11]

Tini picked up her case and bundle of bedding and, with a last look around the apartment, closed the door and locked it. Wickerl Helmhacker was waiting on the landing. Tini handed him the key and turned away. The two women's slow footsteps echoed mournfully in the stairwell as they descended.

Escorted by policemen, they crossed the marketplace, conscious of eyes on them. Everyone knew what was being done with them. For months, periodic batches of Jewish deportees had been leaving, hundreds of people at a time; nobody knew quite where their destination was, other than that it lay somewhere in the vast, vague regions of the Ostland.[12] No news ever came back, and neither did any of the settlers; presumably they were too busy making new lives for themselves in the land that the Reich had set aside for them.

After passing through the market, Tini and Herta were led to the local elementary school. The paving stones in these streets were as familiar to Herta as the soles of her own feet. All the local children had attended the Sperlschule: Edith, Fritz, Kurt, and Herta herself had spent a large part of their lives in its halls and classrooms. It had no pupils now—the SS had closed it down in 1941 and turned it into a holding center for deportations.

They passed through the guarded gateway and along the cobbled alleyway between the tall buildings. The school consisted of a cluster of four-story buildings set back from the street, surrounding an L-shaped schoolyard. Where children had once run and played, SS guards now stood sentry. Trucks were parked, loaded with crates and bundles. Tini and Herta presented their papers and were taken into a building.

The classrooms had been converted into makeshift dormitories, filled with people. In all, the deportees numbered just over a thousand. Everywhere were the faces of friends, acquaintances, neighbors, as well as strangers from more distant parts of the district. Nearly all were women, children, and men over forty. Most young men had gone to the camps, and elderly people over sixty-five were slated for separate deportation to the ghetto at Theresienstadt.*

Tini and Herta were put in a room and left to join in with its little community. News was exchanged, grapevine gossip, enquiries about relatives and mutual friends. The news was almost never good. Their resettlement had been presented to

---

* Now Terezín, Czech Republic.

them as an opportunity for a new life, but Tini hated the idea of being taken from her native city, and was innately suspicious of the future. She had always expected the worst from the Nazis, and they hadn't proved her wrong so far.

In her letter to Fritz and Gustav she had only been able to tell them the bare, devastating fact of their selection. But fearing the worst, she had given a few personal effects to a non-Jewish relative, including her last photograph of Fritz—the one taken in Buchenwald—and had given her sister Jenni a package of clothing to send him. Jenni was in as precarious a position as Tini herself, but so far she'd been omitted from the deportations.[13] The same was true of their widowed older sister, Bertha.[14]

Tini and Herta had been in the holding center for a day or two when the deportees were alerted for departure.[15] Everyone was ordered out into the yard. People crowded the corridors, spilling out through the doors, all carrying luggage and bundles, some burdened with tools and equipment. Their identity cards were inspected, each one stamped *Evakuiert am 9. Juni 1942*,* and they climbed aboard the waiting trucks.

The convoy passed down Taborstrasse and along the broad avenue beside the Danube Canal. Herta looked down at the water, gleaming under the summer sun; come the weekend it would teem with pleasure boats and swimmers. She recalled the time she and her papa had challenged each other to a swimming race, in imitation of Fritz and his friends. Her beloved papa, so gentle and warm. Those had been good days, with summer picnics under the trees by the water. Sometimes her mother, who loved to row, would take the kids out in a boat. It was like a dream now, vivid but remote. Jews had long been barred from the Danube Canal and its verdant banks.

After crossing the canal, the convoy rumbled on through

---

* Evacuated on June 9, 1942.

the streets, eventually pulling in at the Aspangbahnhof, the railway station serving the southern half of the city. A small crowd had gathered around the entrance, held in check by dozens of police and SS troopers. Some were friends and relatives hoping for a last glimpse of their dear ones; others were there simply to gawp at the Jews being herded off like cattle. Tini and Herta helped each other down from their truck and joined the crush as it slowly poured in through the doors into the gloom of the station interior.

Everyone knew of the awful freight wagons in which their menfolk had been taken away to the camps, so it was heartening to find a train waiting for them at the platform, made up of passenger carriages in the attractive cream-and-crimson livery of the Deutsche Reichsbahn. Well now, they thought, this didn't look so bad.

They were ordered to load their luggage into a wagon at the rear of the train; food supplies and medicines had already been stowed. It was a long, slow process. Eventually there came a loud whistle and a voice boomed: "One hour to departure!"[16] The announcement was repeated all along the platform, and people began hurrying in all directions.

Tini, holding tightly on to Herta, pushed through the milling crowd to their assigned place, where a carriage supervisor equipped with a list and an air of flustered importance was marshaling his charges. He was a Jewish official appointed by the IKG, not a police officer or SS, and his presence was reassuring. The sixty or so people assigned to his carriage gathered about him. Tini recognized Ida Klap, an elderly lady from Im Werd, all alone; and a woman about Tini's own age from Leopoldsgasse, also alone; many of the women were unaccompanied, their husbands and sons having been taken and their children—the fortunate ones—having been sent to England or America. Some little ones remained, however. A woman Tini didn't recognize, aged sixty or so, was traveling with three boys and a girl, evidently her grand-children; the youngest, a boy named Otto, was about Kurt's age, and the eldest was a girl of about sixteen.[17] Around them, gray-bearded men in rumpled hats, fellows with pouchy cheeks and jowls, neat, careworn wives in headscarves mingled with young women whose faces were prematurely lined and the disoriented children, some as young as five, staring about in wonder and con-fusion. The carriage supervisor called their names from his list, checking them off against their transport numbers.

"One-two-five: Klein, Nathan Israel!"

A man in his sixties held up his hand. "Here."

"One-two-six: Klein, Rosa Sara!" His wife answered.

"Six-four-two: Kleinmann, Herta Sara!" Herta raised her hand.

"Six-four-one: Kleinmann, Tini Sara!"

The list went on: Klinger, Adolf Israel; Klinger, Amalie Sara . . . Along the length of the platform the fifteen other car-riage supervisors called the rolls of their own sections of the list of 1,006 souls who were going on the journey.

At last their destination was revealed: the city of Minsk. There they would either join the ghetto and work in the various local industries, or farm the land, depending on their skills.

When the supervisors were satisfied that nobody was missing, the evacuees were finally allowed to board, with stern instruc-tions that they were to do so in silence, and keep to their desig-nated seats. The carriages were second class and divided into compartments—comfortable enough, if a little overcrowded. As Tini and Herta took their seats, it was almost like the old days. For a long while now it had been illegal for Jews to venture outside their districts, let alone leave Vienna. It would be inter-esting to see a little of the outside world again.

Smoke and steam poured across the platform, axles squealing as the long train began to move, slowly snaking out of the station, heading north through the city. It crossed the Danube Canal and rolled over the bridge at the west end of the Prater, past the Praterstern and the street where Tini had been born,[18] and a few moments later passed through the northern railway station. This would have been a more convenient place for the Jews of Leopoldstadt to depart from, but the Aspangbahnhof was more discreet.[19] A few minutes later the broad River Danube passed beneath the compartment window, then the final suburbs and the rolling farmland northeast of Vienna.

Though the train stopped occasionally at stations, the evacuees were forbidden to get off. The hours of the long June day dragged by. People read, talked, slept in their seats. Children grew restless and fretful, or catatonic with exhaustion, staring. At regular intervals the carriage supervisor came along and peered into each compartment to check on his charges. A doctor—also appointed by the IKG—was on hand if anybody felt unwell. It was a long time since any Jew had been looked after so solicitously.

They passed through the former Czechoslovakia and entered the land that had been Poland. It was all Germany now. To Tini and Herta the countryside was of particular interest; Gustav had been born in this region during the great days of the Austro-Hungarian Empire, when the Jews had enjoyed a golden age of emancipation. Tini had experienced that era in Vienna, while Gustav had spent his childhood in this beautiful landscape, in a little village called Zabłocie bei Saybusch,* that stood by a lake at the foot of the mountains. The train didn't go there, but it passed nearby, through places that Gustav himself would have recognized, not just from his childhood but from his military service in the war, when he had fought for these same fields and towns against the army of the Russian tsar.

* Now Zabłocie in Żywiec, Poland.

The train also passed near another small town, about thirty miles north of Zabłocie, named Oświęcim. The Germans called it Auschwitz, and had lately established a new concentration camp there. The Vienna train chugged in a wide arc to the west, then resumed its northeastward route, turning away from the setting sun.[20]

There followed a night of ceaseless movement and comfortless dozing, with aching backs and dead limbs. The next morning they passed through the city of Warsaw. Beyond Bialystok they crossed the border, leaving Greater Germany behind and entering the Reichskommissariat Ostland, formerly part of the Soviet Union. About twenty miles farther on, the train reached the small city of Volkovysk.*

Here it stopped.

For a while it seemed no different from any previous halt. Tini and Herta, like everyone else, peered through the window, wondering where they were. The carriage supervisor looked in on the compartment, then moved off. Somehow there was a sense that something wasn't quite right. There was a sound of raised voices at the far end of the corridor, carriage doors opening and heavy boots coming briskly along from both ends. Suddenly, armed SS troopers appeared at the compartment door, and it was flung open.

"Out! Out!" they barked. "All out now!" Shocked and confused, the evacuees scrambled to their feet, grabbing for their belongings, mothers and grandmothers clutching their children. The SS troopers lashed out. "Come on, Jew-pigs! Out now!" Tini and Herta found themselves in the corridor, crushed by people hastening to get to the doors. Any who were slow were kicked or thumped with rifle butts. They poured on to the platform, where more SS troopers were standing by.

The troopers were like none that Tini had ever seen in Vienna;

* Now Vawkavysk, Belarus.

these were Waffen-SS, fighting troops, fiercer and with the Death's Head insignia of the concentration camp division on their collars.[21] They were accompanied by men in the uniforms of the dreaded Sipo-SD, the Nazi security police.[22] They yelled and cursed at the Jews, driving them along the platform—men and women, elderly and children; those who stumbled or fell or who couldn't go fast enough were kicked and beaten, some so badly that their unconscious bodies were left lying on the ground.[23]

They were herded onto another train, this one made up of freight wagons. Into these they were driven at gunpoint, crushed in with scarcely room to move. Then the doors slammed. Tini and Herta, clinging to each other, found themselves in a darkness filled with sobbing, the moans of the injured, praying, and the crying of terrified children. Outside they could hear wagon doors grinding shut all along the train.

After the last door had slammed, they were left in darkness, not moving. Hours dragged by. A few people, broken by the sudden, violent shock, lost their reason during that awful night; they screamed and raved. The SS hauled out the mad and the sick and put them all together in a separate wagon, where they suffered a special hell almost beyond imagining.

The next day, the train began to move. It went painfully slowly. The transport was no longer behind a speedy Reichsbahn locomotive but a plodder from the Main Railway Administration, which served the eastern territories. Since leaving Vienna they had traveled more than six hundred miles in two days; now it took a further two days to cover a quarter of that distance.[24]

Eventually the train came to a halt. Sounds from outside suggested that they were in some kind of station. The terrified people waited for the doors to open, but they didn't. Night came and passed in fear and hunger. Then another day and night. The train sat unattended except for periodic inspections by the Sipo-SD guards. It had arrived on a Saturday, and the

German railway workers in Minsk had recently been awarded the right not to work weekends.[25]

Cramped together in darkness, illuminated only by tiny cracks of daylight in the wagon walls, frightened, with little or nothing to eat or drink, and only a bucket in the corner as a toilet, the deportees endured the dragging hours in horrible uncertainty. Had the plan for them changed? Had they been tricked? On the morning of the fifth day since leaving the comfort of the passenger train, the imprisoned were jolted from their stupor; the train was moving again. Dear God, would this never end?

"Please, dear child," Tini had written to Kurt, almost a year ago now, "pray that we are all reunited in good health." She had never quite let go of that hope. "Papa wrote . . . thank God he is healthy . . . the knowledge that you are well taken care of by your uncle is his only joy . . . Please, Kurtl, be a good boy . . . I hope they have good things to say about you, that you keep your things and your bed in order and that you are nice . . . You have a wonderful summer, soon the beautiful days will be over . . . All the kids here envy you. They don't even get to see a garden."[26]

With a shrieking of steel on steel and a thump and rattle of wagons bumping, the train halted again. There was silence, and then the wagon door slammed wide open, flooding the imprisoned with blinding light.

✡

Precisely what befell Tini and Herta Kleinmann that day will never be known. What they witnessed, what they did or said or felt was never recorded. Not a single one of the 1,006 Jewish women, children, and men brought to the freight yard at Minsk railway station on the morning of Monday, June 15, 1942, was ever seen again or left any account.

But general records were kept, and there were other

transports from Vienna to Minsk during that summer from which a handful of individuals brought back their stories.[27]

When the wagon doors opened, the people inside—bruised, bone-weary, aching, starving, dehydrated—were ordered out. They were pushed around, scrutinized by Sipo-SD men, and quizzed about their trade skills. An officer addressed them, reiterating what they had been told back in Vienna—that they would be put to work in industry or farming. Most of them, unable to do without hope, were reassured by this speech. A few dozen of the healthier-looking adults and older children were selected and taken aside. The remaining multitude were herded to the station barrier, where their belongings were taken from them. The wagonloads of luggage, food, and supplies that had been brought from Vienna were also seized.[28] Waiting outside the station were trucks and closed vans, into which the people were loaded.

The convoy drove out of the city, heading southeast into the Belarusian countryside—a vast, flat plain of field and forest, dusty under a huge sky.

When the German forces took this land from the Soviet Union the previous summer, they had rolled through it like a consuming wave. Immediately in their wake had come a second wave: Einsatzgruppe (Task Force) B, one of seven such units deployed behind the front lines. Commanded by SS-General Arthur Nebe, Einsatzgruppe B comprised about a thousand men—mostly drawn from the Sipo-SD and other police branches—divided into smaller subunits, or Einsatzkommandos. Their role was to locate and exterminate all Jews in captured towns and villages, a task in which they were often willingly assisted by units of the Waffen-SS and Wehrmacht, and in some areas, such as Poland and Latvia, by local police.[29]

Not all Jews were murdered immediately. That was impracticable, given the millions who inhabited these regions. Besides, the Nazis had learned in Poland how to make Jews contribute to the war economy. A ghetto was established in Minsk, and its industry made to serve the Reich and line the pockets of corrupt officials. Now, with the implementation of the Final Solution, Minsk had been chosen as one of its principal centers.

The task of organization fell to the local Sipo-SD commander, SS-Lieutenant Colonel Eduard Strauch, a veteran Einsatzgruppe officer. He surveyed the area and chose to establish a concentration camp at the secluded little hamlet of Maly Trostinets, a former Soviet collective farm about seven miles southeast of Minsk. The camp was small, never intended to hold more than about six hundred prisoners to work the farmland and provide a Sonderkommando* for its main purpose, which was mass murder.[30]

Of the tens of thousands of people—mostly Jews—brought to Maly Trostinets, few ever saw the camp itself. After the Sipo-SD had selected a handful from each transport for the labor force, the trucks carrying the remaining hundreds drove out in the direction of Maly Trostinets. Along the way, they would stop off at a meadow outside the city. Sometimes the selection for the camp would be made here if it hadn't already been done at Minsk railway station.[31] From the meadow, at intervals of an hour or so, individual trucks would drive on while the rest waited.

The trucks drove to a half-grown pine plantation about two miles from the camp. There, one of two possible fates awaited the captives. For the majority it was quick, for some slower. But the end was the same. There was a clearing among the trees where a huge pit had been excavated by a Sonderkommando, about two hundred feet long and ten feet deep. Waiting beside it was a platoon of Waffen-SS under SS-Lieutenant Arlt. Each man was armed with a pistol and twenty-five rounds of ammunition; more boxes of cartridges were stacked nearby.[32] About

---

* Special labor detail: concentration camp prisoners forced to handle victims before and after executions.

six hundred feet out from the clearing, a cordon of sentries from a Latvian police unit stood guard, to prevent any victims escaping or any potential witnesses venturing near.[33]

Disembarked from the truck, the women, men, and children were forced to strip to their underwear, leaving behind any possessions they had on them. At gunpoint, in groups of about twenty, they were marched to the edge of the pit, where they had to stand in a line, facing the edge. Behind each person stood an SS trooper. On the order, the victims were shot in the back of the neck at point-blank range, and fell into the pit. Then came the next batch. When they had all been shot, a machine gun mounted at the end of the pit opened fire on any corpses that seemed to be still moving.[34] After a short interval, the next truck would arrive, and the process would be repeated.

What made those people submit? From the first who faced the empty pit to those who saw it already half-filled with the corpses of their neighbors and friends, and heard the shots being fired—what enabled them to walk into place, stand, and be shot down? Were they subdued by terror? Had they resigned themselves to their fate, or suffered an existential self-negation? Or did they still retain, until the very last split second with the pistol at their neck, a hope that the shot would not fire, that somehow they would be reprieved? A few did try to run, although they didn't get far, but overwhelmingly the victims went quietly to their deaths.

At Maly Trostinets there was none of the undisciplined fury and euphoria that had often characterized Einsatzgruppe killings elsewhere, in which infants had their backs broken and were hurled into the pits, and the murderers laughed and raged as they killed. Here it was just cold, clockwork execution.

And yet it told on the killers' minds. Even these men had consciences of a sort—wizened, stunted consciences, just enough to be rubbed raw by the endless blood and guilt. Arlt's men were provided with vodka to numb the feeling,[35] but it didn't heal the damage. For this reason the SS had experimented with alternative methods that would allow them to exterminate but avoid bloodying their hands. This had brought about the second, slower, method of execution employed simultaneously at Maly Trostinets.

At the beginning of June, mobile gas vans had been introduced. There were three of them—two converted from Diamond goods vans, and one larger Staurer furniture removal van. The Germans called them *S-Wagen*, but the local Belarusian people called them *dukgubki*—soul suffocators.[36] While the majority of Jews were shot at the pit, some—probably two or three hundred from each transport—went in the vans. The lottery happened at the station in Minsk, where some were loaded into the regular trucks, and some into the *S-Wagen*, crammed in so tightly that they crushed and trampled one another.

Once the shootings had been completed, the gas vans started up and drove to the plantation, where they parked beside the corpse-filled pit. Each driver or his assistant connected a pipe from the exhaust to the van interior, which was lined with steel. Then the engine was started. The people trapped inside immediately began to panic; the vans shook and rocked on their suspension with the violence of their struggle, and there were muffled sounds of screaming and hammering on the sides. Gradually, over the course of about fifteen minutes, the noise and shuddering lessened and the vans grew still.[37]

When all was quiet, each van was opened. Some of the bodies, which had piled up against the door, fell out onto the ground. A Sonderkommando of Jewish prisoners climbed up inside and began hauling out the rest of the corpses, heaving them into the pit. The van interior was a scene of unbearable horror; the bodies were streaked with blood, vomit, and feces; the floor was littered with broken spectacles, tufts of hair, and even teeth lying in the mess, where the victims had fought and clawed the people near them in their demented efforts to escape.

Before the vans could be used again, they were taken to a pond near the camp and the interiors thoroughly sluiced out. The delay this caused, together with the small number of vans available and frequent mechanical failures, was the reason why firing squads were still used. The SS was still working to refine its methods of mass murder.

SS-Lieutenant Arlt wrote in his log for that day: "On 15/6 there arrived another transport of 1,000 Jews from Vienna."[38] That was all. He had no interest in describing what he and his men had done; it was just another day's work, over which the SS felt it was best to draw a veil of discretion.

אמא

A summer sun lay hot and lazy on the slow-moving surface of the Danube Canal. The faint, delighted squeals of children drifted over the water from the grassy banks where families sat with picnics or strolled under the trees. Pleasure boats cruised and row boats scudded across the expanse between them.

It was all far away from Tini's senses as she pulled on the oars—a pleasant, distant background music of laughter. Sunlight sparkled on the splashes with each lift of the oar blades from the water, illuminating the faces of her children. Edith, smiling serenely, Fritz and Herta still little kids, and Kurt, the last-born and beloved, a tiny speck scarcely out of diapers. Tini smiled and heaved at the oars, sending the boat surging across the water.[39] She was a good rower—had been since her girlhood. And she doted on her family; at the age of twelve she had been made a counselor to the younger schoolchildren because she loved it so much; to nurture and succor was part of her makeup, and in motherhood it had its purest expression.

The sounds of the other boats and the revels on the far banks faded, as if a mist had descended, closing off the boat from the world. The oars dipped and splashed, and the boat glided on.

In a drawer in a chest in faraway Massachusetts, Tini's last letters to Kurt lay gathered. The German in which they were written was already leaking away from his comprehension as his child's mind adapted to his new world. He had absorbed her meaning, but was already slowly, insensibly beginning to forget how to read her words.

My beloved Kurtl . . . I am so happy that you are doing well . . . write often . . . Herta is always thinking of you . . . I am afraid every day . . . Herta sends hugs and kisses. A thousand kisses from your Mama. I love you.

✡

That night, after the Sonderkommando had backfilled the pit, dusk fell on the silent clearing among the young pines. Birds returned, night creatures foraged among the weeds and ran over the disturbed soil of the pit. Beneath lay the remains of nine hundred souls who had boarded the train in Vienna: Rosa Kerbel and her four grandchildren—Otto, Kurt, Helene, and Heinrich—and the elderly Adolf and Amalie Klinger, five-year-old Alice Baron, the spinster sisters Johanna and Flora Kaufmann, Adolf and Witie Aptowitzer from Im Werd, Tini Kleinmann and her pretty twenty-year-old daughter, Herta.

They had believed that they were going to eke out a new life in the Ostland, and that perhaps one day they would be reunited with their dear ones—husbands, sons, brothers, daughters—who had been scattered to the camps and far countries.[40] Beyond all reason, beyond all human feeling, the world—not only the Nazis but the politicians, people, and newspapermen of London, New York, Chicago, and Washington—had closed off that future and irrevocably sealed it shut.

10. A Journey to Death

אבא

The summer sun was lowering, spreading an orange cast over the branches and long, coal-gray shadows across the forest floor. Gustav's ears were filled with the rasp of saws on tree trunks and the urgent grunts of men, the pumping of his own blood, and the heave of his breath as he and his workmates hoisted a tree trunk onto their wagon.

In a way it was pleasant to be out in the woods again, away from the grit and dust and mud, but the kapo, a vindictive sadist called Jacob Ganzer, was a hard driver. "Faster, pigs! You think those logs will stack themselves? Move!"

At such a speed the work was not only exhausting but dangerous. Gustav and his workmates raised the massive log and launched it onto the stack atop the creaking wagon. Not a second to spare to catch their breath or ensure that the stack was stable—another log was ready to be heaved up, and Ganzer was barking furiously. Gustav took one end of the massive trunk, his mate, a prisoner named Friedmann, applied his shoulder to it, and other hands took up the weight; muscles cracking, they strained it upward, over the sideboard, up toward a space on the pile. With Ganzer's hectoring in their ears, somebody let go before the trunk was settled. It rolled back, an unstoppable mass weighing hundreds of pounds, bringing others with it.

The trunk rolled over Gustav's hand; his brain scarcely had time to feel the cracking pain in his fingers before it slammed into his body and Friedmann's, knocking them to the ground and landing on top of them.[1]

Gustav lay pinned like a butterfly on a card, staring up at the swirling canopy of leaves flickering in the evening sun, his body a mass of pain, his ears filled with screams and groans and shouting. Then striped uniforms were in his vision, hands scrabbling at the trunk, lifting it off him, but he still couldn't move. Looking around, he saw men picking themselves up, with bloodied hands and faces, others sprawled and moaning. Friedmann lay a few feet away, motionless, whimpering hoarsely; he had taken most of the force of the falling trunk on his chest. Blood was oozing from his mouth.

Hands clasped Gustav's body, and he was picked up and carried from the clearing. Through the pain, he saw the trees flit by, the sky fading and tilting, heard the grunting of the men bearing him. The gatehouse passed by him, and then he was entering the infirmary and being laid on a pallet.[2]

Seven other men from his detail came after him, either carried or hobbling by themselves. Friedmann arrived last on a stretcher. He couldn't move; his ribcage was crushed and his spine broken. He lay in helpless agony.

Gustav's chest had taken some of the impact, and his broken fingers were on fire with pain. The lottery had finally run against him, as it did for everyone. It could only run true so many times, and the longer one was forced to take part, the more certain it was that it would turn bad. The prospects for a badly injured man were grim. The doctor's needle and a vein full of phenol or hexobarbital was the likely fate, and then— smoke from the crematorium chimney.

Friedmann died mercifully quickly from his injuries. Most of the other men, lightly injured, were out of the infirmary in a short time. But Gustav remained. The days dragged by, and he was placed in a small ward adjoining Operating Room II. If he didn't know already what this meant, he would quickly learn; Operating Room II was where lethal injections were given, and this ward was its waiting room.[3]

For a while Gustav was left alone; periodically a sick or badly injured man would be selected and taken through to OR II. They never came back. The doctor looked at Gustav each time and passed him by; he was too badly injured to bother with. It would be a waste of drugs to inject a prisoner who would certainly die soon of his own accord. The doctor knew nothing of the will and resilience of Gustav Kleinmann.

There was a friendly orderly called Helmut, who carefully tended Gustav when the doctor wasn't around, and he managed to cling on determinedly to life, racked by pain day and night. It slowly subsided, and after six weeks he had recovered enough to be discharged. He was still on a knife's edge; lacking the strength to return to the haulage detail or even the infirmary wagon, he was now a useless mouth, liable to be sent back to OR II for liquidation.

His friends and his trade skills saved his life. Words were exchanged between friendly kapos, and Gustav was transferred to the DAW factory, which manufactured military supplies like cartridge cases, barrack-room lockers, and aircraft parts, and converted trucks into mobile canteens.[4] Gustav was given work as a saddler and upholsterer. He began to convalesce.

For the first time since his arrival in the camp—almost the first time since the Anschluss—Gustav was able to practice his proper trade again. He was happy—or as happy as one could be. The work was congenial, and he made good friends. His foreman was a German political prisoner named Peter Kersten, a former Communist Party city councilor—"a very brave man," Gustav thought, "I get along with him very well." He even managed to wangle a work placement for a Viennese friend, Fredl Lustig, a workmate from the haulage column. Together they made a contented little band.

So it went on until the beginning of October. And then, like a nightmare resuming after a moment's gasping wakefulness, everything suddenly, calamitously changed.

בן

Fritz and his workmate lifted a heavy concrete lintel from the scaffold bed and carefully eased it into place in the wall above the window space. Fritz positioned it, checking its level and fit.

Over the past couple of years, his skills as a builder had developed under Robert Siewert's tutelage. He'd mastered all kinds of brickwork and stonework, plastering and general construction. The Siewert detachment was now hard at work on the site of the new Gustloff Werke, a large factory being built beside the Blood Road, opposite the SS garage complex. Once completed, it would turn out barrels for tank and antiaircraft guns, as well as other armaments. Most of the exterior walls were already up, and Fritz had been put to work on the huge factory windows. He was expected to complete two a day, constructing the fenestrations, setting the lintels and fixing them in place, a job requiring a highly skilled bricklayer and a great deal of care.

His workmate, Max Umschweif, was a relative newcomer to Buchenwald, having arrived the previous summer. A slightly built Viennese with the face of an intellectual, he had fought with the International Brigade against the fascists in Spain. After the defeat, he and his comrades had been interned in France; returning to Vienna in 1940, he'd been arrested by the Gestapo as a known antifascist. Fritz loved hearing his stories about the war in Spain, but was utterly bewildered that he had voluntarily returned to Austria knowing that the Gestapo would be after him.

Tapping the lintel into its final position with the butt of his trowel, Fritz checked it with a spirit level, then quickly and skillfully mortared it in place. It was pleasant working up here on the scaffolding. While the SS overseers constantly harassed and beat the brick and mortar carriers, they never ventured up the ladders onto the scaffolding. Satisfied with the lintel, Fritz turned and took a moment to stretch his muscles. There was a fine view over the forest from up here: the oaks and beeches were beautiful in their October glory, dappled with gold and shades of copper. Far away, the spread of Weimar could be seen, and the rolling farmlands around.

Fritz had been through some terrible experiences in recent months—the departure of Leo Moses, his father's near death, close friends murdered by the SS. And yet the worst thing was the news about his mother and Herta, and the agony of not knowing what had become of them.

His reverie was interrupted by a call from below. "Fritz Kleinmann, come down here!" He clambered down the ladder and found one of the laborers waiting for him. "Kapo wants you."

He went in search of Robert Siewert, and found him wearing the grave look Fritz had seen before. Siewert took him quietly to one side and put his arm round his shoulders, pulling him close as if he were his own son; he'd never done such a thing before, and Fritz guessed that bad news was coming. "There is a list in the records office of Jews to be transferred to Auschwitz," Siewert said simply. "Your father's name is on it."

The shock was beyond anything Fritz had ever felt. Everyone knew the name of Auschwitz, one of the crop of camps the SS had been establishing in occupied countries. All year there had been talk in Buchenwald: rumor and news from far away, as well as events witnessed in the camp itself, indicated that the drama of the Jews was entering its last act, that the Nazis meant to finally dispose of those who had not emigrated or died already. Since the spring there had been disturbing whispers about special gas chambers being built in some camps, in which hundreds of people at a time could be put to death. One such camp was Auschwitz. A transfer there meant only one thing.

Siewert explained what he had learned. The list was a long one, comprising almost all the Jews still alive in Buchenwald; the only exceptions were those like Fritz who were required for the construction of the Gustloff factory.

Fritz was dazed and appalled; he knew so many youngsters in the camp who had lost their fathers, and it had been his abiding fear that he would become one of them. "You will have to be very brave," Siewert said.

"But Papa does useful work in the factory," Fritz objected.

Siewert shook his head. Factory work was meaningless. "It is *everyone*," he said. "Every Jew except builders and bricklayers is going to Auschwitz. You are one of the fortunate ones." He looked Fritz in the eye. "If you want to go on living, you have to forget your father."

Fritz struggled to find words. "That's impossible," he said. With that, he turned on his heel, scrambled back up the ladder to the scaffold, and went back to work.

✡

Just over four hundred names were on the list drawn up by Buchenwald's headquarters. A few days earlier they had received an order sent on behalf of Himmler to all camp commandants: on the wishes of the Führer himself, all concentration camps located on German home soil were to be made Jew-free. All Jewish prisoners were to be transferred to camps in former Polish territory—namely Auschwitz and Majdanek.[5]

In Buchenwald there were only 639 Jews left alive: those who had survived the random murders, transfers, and euthanasia transports. Of that total, 234 were employed in building the factory; they were to be retained for the time being, while the rest were slated for Auschwitz.[6]

In the evening of Thursday, October 15, a few days after Fritz's conversation with Robert Siewert, all the Jewish prisoners were ordered to assemble in the roll-call square.[7]

They knew what to expect, and it was exactly as Siewert had foretold. Fritz heard his number called out among the roll of skilled construction workers. These men were ordered to return to their barrack blocks. Leaving his papa behind, Fritz marched off with his workmates, his insides knotted with dread and indignation.

Gustav and the rest of the four hundred were informed that they were to be transferred to another camp. From this moment on, they would remain in isolation. They were marched to block 11, which had been cleared to make way for them, and shut inside, barred from all contact with other prisoners. There they waited for the transfer to begin.

אב ובן

That evening, Fritz couldn't settle, couldn't clear his mind of the image of his papa left standing among the condemned men. The prospect of being parted forever was unbearable. All night it tormented him. Fritz knew that Robert Siewert's words were wise and sensible and kindly meant: he had to learn to forget his papa if he wanted to survive. But Fritz could not imagine himself being able to keep on living if that was what it took. His fears about his mother and Herta had planted a feeling of despair in him, and he didn't see how he could possibly live if his papa were murdered.

In the early hours a rumor passed among Fritz's block-mates: three of the prisoners in block 11 had been taken to the infirmary during the night and killed by lethal injections. The rumor was false, but it helped push Fritz toward an ultimate resolution.

The next morning, before roll call, he sought out Robert Siewert and pleaded with him. "You have contacts," he said. "You have friends who are clerks in the headquarters office." Siewert nodded; of course he did. "I need you to pull whatever strings you can to get me on that Auschwitz transfer."

Siewert was aghast. "What you're asking is suicide. I told you, you have to forget your father," he said. "These men will all be gassed."

But Fritz was adamant. "I want to be with my papa, no matter what happens. I can't go on living without him."

Siewert tried to dissuade him, but the boy was immovable. As roll call was ending, Siewert went and spoke to SS-Lieutenant Max Schobert, the deputy commandant. While the prisoners began forming up to march off for morning work, the call went up: "Prisoner 7290 to the gate!"

Fritz reported to Schobert, who asked him what was the matter. This was the moment of no return. Steeling himself, Fritz explained that he couldn't bear to be parted from his father, and requested formally that he be sent to Auschwitz with him.

Schobert shrugged; it was all the same to him how many Jews were sent to be exterminated, and he granted the request.

With a word, Fritz had done the unthinkable, stepping voluntarily from the roll of the saved to that of the condemned. He was placed under guard and led back across the square to block 11. The door opened and he was pushed inside.

The barrack, built for only a couple of hundred men, was full to bursting. Fritz found himself looking into a mass of striped uniforms, standing, sitting on the few chairs or squatting on the floor, craning at the windows to see what was happening outside. Dozens of faces turned to stare at Fritz as the door slammed shut. Almost every one of them was an old friend or a mentor—the thin, bespectacled face of Stefan Heymann, perpetually surprised, now astonished; his friend Gustl Herzog; the courageous Austrian antifascist Erich Eisler and Bavarian Fritz Sondheim . . . the astonishment on their faces gave way to horror when they learned why he was here. They protested and implored, just as Siewert had, but Fritz pushed past them, looking for his papa . . .

. . . and there he was, among the crowd, the familiar lean, lined face with its calm, gentle eyes. They rushed to each other and embraced, both sobbing with joy.

Later that night, Robert Siewert came to talk to Fritz; he was required to sign a paper acknowledging that he was going on the transport of his own free will.[8] Their parting was painful; Fritz owed Siewert his position, his skills, his very survival during the past two years.

On the morning of Saturday, October 17, after two days in suspense, the 405 Jewish transferees—Polish, Czech, Austrian, and German—were informed that they were to be transported that day. They were ordered to take no possessions with them. They were issued a meager ration of food to take with them on their journey—Gustav's consisted of a single hunk of bread—and then led outside.

The mood in the camp was unusually somber, even among the SS. Previous transfers had been marched out under a hail of abuse from the guards, but the four hundred Jews marched to the gate in silence. It was as if they all recognized that this was different, a momentous thing that was not to be treated lightly.

Outside the gate, a convoy of buses awaited them. Fritz and

Gustav sat in civilized comfort as they drove down the Blood Road, along which they had run in terror three years, two weeks, and one day earlier. How much they had changed since then; how much they had seen. At Weimar station, they were loaded into cattle wagons—forty men in each.[9] Extra boards had been nailed on to close up gaps and make the wagons absolutely secure.

As it set off, the mood in Fritz and Gustav's wagon—which they shared with Stefan Heymann, Gustl Herzog, and many other friends—was depressed. In the daylight leaking through cracks in the wagon walls, Gustav took out his diary, keeping it out of view of the others. Having been forewarned of the transfer, he'd ensured that he had it concealed under his clothing when they were moved to the isolation block. This battered little notebook had come to represent his grip on sanity, his record of the reality of life now, and he wouldn't wish to be parted from it. But so long as he was with Fritz, he felt he could face anything.

"Everyone is saying it is a journey to death,"[10] he wrote, "but Fritzl and I do not let our heads hang down. I tell myself that a man can only die once."

PART III
Auschwitz

11. A Town Called Oświęcim

אחים

Another train, another time . . .

Gustav woke from a doze with sunlight rippling across his eyelids and his nostrils filled with the odors of serge, sweaty male bodies, tobacco smoke, leather, and gun oil. His ears were filled with the steady clatter of the train and the mutter of men's voices, suddenly raised in song. The men were in good spirits, even though they might be going to their deaths. Gustav rubbed his neck, sore where he'd rested his head on his pack, and retrieved his rifle, which had slipped to the floor.

Standing up and peering through the side slot, he felt the warm summer wind on his face and smelled the scents of meadows fleetingly through the smoke from the locomotive. The rolling wheat fields were at the green-gold stage, ripening toward the harvest. A church spire broke through a gap in the distant rise; behind stood the green Beskid Mountains, and beyond them the ghostly curtain of the Babia Góra, the Witches' Mountain. This was the land of his childhood. After six years in Vienna it looked strange, in that peculiar way of a vivid memory suddenly unearthed.

He'd been drafted into the Austrian Imperial and Royal Army in the spring of 1912, the year he turned twenty-one.[1] As a born Galician he'd been placed in the 56th Infantry Regiment, based in the Cracow district.* For most young working-class men,

* Now Kraków, Poland.

army service was a welcome interlude: conditions were good, and it opened their horizons. Many were illiterate, low-paid workers; most had never been farther than the next village. In Galicia the majority didn't even speak German; many couldn't even tell the time.[2] Gustav had seen more of the world than most of his fellow recruits, having lived in Vienna, and he spoke both Polish and German; but as an apprentice upholsterer he was poor, and the army provided some stability. It was an exciting environment—Austria's empire had once been the greatest in Europe, and the army still preserved its imperial panoply of hussars and dragoons, colorful, dashing dress uniforms and endless pomp with the flags and banners of the imperial Double Eagle fluttering over it all.

For Gustav military service had meant a return to his homeland, spending most of the first two years in a garrison town north of the Beskid Mountains, about halfway between his home village of Zabłocie and a town called Oświęcim, a pretty, prosperous, but otherwise unremarkable place on the Prussian border. For two years he lived barrack life: parades, bootblacking, and brass polishing, with occasional field exercises and maneuvers. And then, in 1914, just when the 1912 intake thought they would soon be done with the army and going back to their farms and workshops with their manhoods made, the war came.

All of a sudden, the 56th Infantry Regiment was mobilized, and marched with the rest of the 12th Infantry Division to the railway station to embark for the fortress town of Przemyśl[3]—the regiment's jumping-off point for the advance into Russian territory.[4] Gustav and his comrades marched with a lively step under their heavy packs as the band blared out the vibrant tune of the "Daun March," immaculate in their gray uniforms with steel-green facings, their moustaches waxed, grinning to the waving girls and as pleased with themselves as only young men can be. They were off to chase the Russians all the way to St. Petersburg.

They had less spring in their step five days later, after a journey in cattle wagons and a long, punishing forced march under fifty-pound packs, winter overcoats strapped on, ammunition, spade, and rations for days, rifle straps chafing and feet sore. Lance Corporal Gustav Kleinmann and his platoon-mates were more ready for bed and bottle than for battle. They got neither that first day. Their objective was the city of Lublin, where they were supposed to link up with a Prussian advance from the north. While regiments on their left flank met heavy Russian resistance and took a lot of casualties, the 56th barely made contact; they just marched and marched all day long, pushing into Russian territory.[5]

בן

Gustav eased his wounded leg into a more comfortable position. Outside, a hard Galician frost bit at the edges of the windowpanes and snow lay heavy on the ground.

After the blazing summer, a terrible autumn and a wretched winter had followed. Despite driving the Russian army back in disarray, the Austrian troops had been let down by poor leadership, and the Germans had failed to support them properly. The Russians had soon rallied and begun recapturing ground.[6] It had turned into a rout, with Austrian regiments breaking and falling back all along the line.

The civilian population panicked, and the railway stations and roads had become choked with refugees. Jews were especially terrified; tsarist Russia's anti-Semitic laws were notorious. Indeed, many Galician Jews were descendants of those who had fled Russian pogroms. The advancing Russians expropriated Jewish property and extorted money from them with threats of violence, Jews were dismissed from public offices, and some were seized as hostages and taken away to

Russia.[7] Refugees flooded west and south toward the heartlands of Austria-Hungary. At first they sought sanctuary in Cracow, but by the autumn even that city was under threat, so the refugees began heading for Vienna. The authorities set up embarkation stations for them at Wadowitz* and Oświęcim.[8]

Eventually Austria's forces—with Gustav and the 56th to the fore—had fought the Russians to a standstill, and the front line settled just short of Cracow. The armies dug trenches and began the dreadful attrition of bombardment, raids, and hopeless attacks. By the new year, Gustav and his comrades—what remained of them—were in the front line outside Gorlice, a town about sixty miles southeast of Cracow. The trench line was little more than a series of shallow ditches protected by a single strand of barbed wire, hemmed in by open ground that was pounded by Russian artillery.[9] The enemy held the town and dominated the ground in front of it from a stronghold in a large hilltop cemetery on the western outskirts.

And there they had sat through the biting winter. For Gustav, it had been a kind of reprieve when he was wounded—a bullet through the left forearm and calf.[10] He'd stayed briefly in the auxiliary hospital at Bielitz-Biała,† a large town close by Zabłocie (he knew the place well, having worked there as a baker's boy in his early teens), and in mid-January he'd been moved here, to the reserve hospital in the next town—the transport hub and army base of Oświęcim, or, as it was called in German: Auschwitz.

Gustav had known the place in childhood. The town itself had been pleasant in peacetime, with fine civic buildings and an ancient, picturesque Jewish quarter that attracted tourists.[11] It stood at the confluence of the Vistula and the Soła, the river that meandered down from the lake by the village where Gustav had been

* Now Wadowice, Poland.
† Now Bielsko-Biała, Poland.

born. The military hospital at Oświęcim was a little way from the town, across the Soła in the outlying hamlet of Zasole—a group of modern barracks standing in neat rows near the riverbank (not an ideal spot: the ground was marshy and in summer plagued with insects). Originally the barracks had been attached to a transit camp for seasonal migrant workers flowing from Galicia into Prussia, but since the outbreak of war the lines of workers' barracks had stood empty.[12]

For Gustav, worse than the ache of the wounds—which were almost healed now—was the wrench of being away from his comrades while they were still in the line. He was determined not to malinger; his wounds weren't debilitating, and despite his slight, even delicate appearance, with his soft eyes and large ears, he had proven a tough young man, with a surprising capacity for taking hardship and injury.

But for now he was at peace here, the only sounds the brisk footsteps of nurses and the low murmur of voices.

אחים

Bullets smacked into sides of the tombs, flinging stone splinters in Gustav's face. He and his men held on and returned fire, pressing ahead, foot by foot, into the cemetery.

Gustav was only a month out of hospital and already back in the thick of it—back to Gorlice, back to the frozen trenches at the foot of the slope below the town, back to the sporadic fall of shells and the steady attrition. Then came this day— February 24, 1915—when the division launched an assault on the heavily defended Russian positions.

To Corporal Gustav Kleinmann's eye it looked like a suicide mission: an uphill frontal attack against a large force in a secure, easily defended position. The cemetery on his company's front was a traditional Catholic one—a city of little tombs of limestone and marble, packed close together. It was a veritable fortress, and Gustav's company had been cut to pieces in the first approach. With their sergeant and platoon officer killed, Gustav and his right-hand man, Lance Corporal Johann Aleksiak, had improvised a plan of their own to avoid wasting any more lives.[13] Leading the remnants of the platoon—now consisting of just themselves, two lance corporals and ten privates—they had skirted to the left flank of the enemy position, where they were sheltered from the Russian fire, and advanced from there. Infiltrating the fringes of the cemetery, they were among the tombs before the Russians were aware of them. Immediately, withering gunfire lashed at them. They returned fire as best they could and pressed on. The Russians began lobbing hand grenades, but still Gustav and his squad pushed forward, driving the enemy back.

They had advanced fifty feet inside the enemy perimeter when the alleys between the tombs became too narrow for effective firing. Gustav stopped his men and ordered them to fix bayonets. With their blood hot and furious, they launched their final, savage assault.

It worked; the Russians were prised out of their positions on the points of the Austrian bayonets. Gustav's flank attack had drawn the main force of the Russian defense, allowing the rest of 3 Company to advance into the cemetery. Between them they took two hundred Russian prisoners that day, part of a total haul of 1,240 captured by the regiment.

In the face of the setbacks the Austrian army had suffered since the start of the war, the capture of Gorlice cemetery was a significant enough achievement to earn a deluge of medals and even a passing mention in a report by Field Marshal von Höfer.[14] Not for the first time, or the last, a knife-edge battle had turned on the initiative of a lowly noncommissioned officer.

בן

Rabbi Frankfurter chanted the last blessings of the Sheva Brachot, the seven blessings of marriage, his voice echoing hauntingly through the synagogue-chapel of Vienna's Rossauer barracks. Beneath the wedding canopy held up by his comrades, Gustav stood in his best dress uniform, the Silver Medal for Bravery 1st Class gleaming on his breast. Beside him was his bride, Tini Rottenstein, radiant, her white lace collar and silk flowers bright against the dark fabric of her coat and broad-brimmed hat.

Two years had passed since that day in Gorlice cemetery. Gustav and Johann Aleksiak had both received the Silver Medal, one of Austria's highest awards. Their commanding officer had described their actions as a "clever, unprecedentedly courageous approach" in which the two corporals had "excellently distinguished themselves."[15] It had been a fierce battle, and over a hundred men of the 56th Infantry Regiment received decorations.[16] Since that day, despite setbacks, the Austrians had driven the tsar's army across the Vistula and out of Galicia, capturing Lemberg,* Warsaw, and Lublin. In August that year Gustav had been wounded again, this time a much more serious injury to the lung.[17] He had recovered eventually and returned again to action.

"May the barren one rejoice and be happy at the gathering in of her children in joy." Rabbi Frankfurter's chanting filled the room. "Blessed are You, Lord, who created joy and happiness, groom and bride, gladness, jubilation, cheer and delight, love, friendship, harmony and fellowship . . . who gladdens the groom with the bride." Then he placed the traditional glass on the floor by Gustav, who brought his

---

* Later Lwów, Poland; now Lviv, Ukraine.

boot heel down and shattered it. *"Mazel tov!"** yelled the congregation.

The rabbi spoke, reminding Tini of the solemnity of wedding a soldier, and touched on the goodness of the Austro-Hungarian Empire to its Jewish people; he likened the new Emperor Karl to the sun shining on the Jews; his forebears had brought down the walls of the old ghettos and "installed Israel" in their realm.[18] Austria had always had its share of anti-Semitism, it was true, but since the emancipation of the Jews under the Habsburg emperors, they had lived well and achieved much. With this foundation, they could make their way with their own hands and hearts.

Gustav and Tini walked out of the synagogue that day into a new era. Gustav wasn't done with fighting; he would see more action on the Italian front and earn more decorations, helping Austria and Germany fight their slow, inevitable, bloody defeat. But he survived in the end and came home to Vienna. In the summer of the first year of peace, Edith was born, the first child of many. The old empire had been broken up by the victorious Allies; Galicia had been ceded to Poland, Hungary was independent, and Austria was reduced to a rump. But Vienna was still Vienna, the civilized heart of Europe, and Gustav had more than earned his family's place in it.

Many didn't see it that way. People in Austria and Germany began telling themselves stories to relieve the shame of losing the war. It was the fault of the Jews, many said; they had thrived in the wartime black market; fingers were pointed at the floods of Jewish refugees fleeing the front, and how they had worsened the food crisis in the cities; tales were told about how Jews had shirked their duty and avoided military service; their pernicious influence in government and commerce had been a knife in

* Congratulations/good luck (Hebrew).

Germany's and Austria's back. In the Vienna parliament there was anti-Jewish agitation from German nationalists and the conservative Christian Social Party, and newspapers began to print dire threats of pogroms.[19]

Yet the promise lived on. The outburst of anti-Semitism settled down to a murmur, and the Jews of Vienna continued to thrive. Gustav sometimes struggled to make a living, but he never despaired, throwing himself into socialist politics in a bid to ensure a brighter future for all working people, and to win prosperity for his children's future.

אבא

Another train, another time, another world . . . and yet the same.

Gustav sat in darkness, rocked by the motion of the train. Around him the air was thick with the familiar stench of unwashed bodies, stale uniforms, and the latrine pail, and alive with the dull murmur of voices. Dozens of men in a space so small that they could scarcely move; getting to the piss bucket in the corner was an ordeal.

Two days had passed since boarding the train at Weimar. Gustav's eyes had adapted to the slivers of light leaking through cracks around the door and gratings, just enough to write a few brief lines in his diary. It must be around noon now; the light was at its brightest, and the faces of his comrades were discernible: Gustl Herzog was there, and the long, earnest features of Stefan Heymann, as well as Gustav's friend Felix "Jupp" Rausch, and Fritz sitting close to some of his young friends, including Paul Grünberg, a Viennese who was the same age as him and had been one of Siewert's apprentices but hadn't completed his training.[20] Without water or blankets they were thirsty and cold, and the mood was profoundly depressed.

He could neither see it nor smell it, but Gustav knew the landscape through which he guessed they were passing now: the fields, the green distant hills and mountains, the quaint little villages. He had grown up here, bled for his country here, and now the rail tracks were bringing him back one last time, to die here.

Behind him, the family he had begun with such hope lay broken and scattered. The promise of 1915, when they pinned the medal to his chest, and of 1917, when he'd ground the glass beneath his heel and joined with Tini in marriage, and the promise of 1919, when he'd held baby Edith for the first time, the promise that Israel had been built in Austria: that promise had been crushed under the wheels of this vast, insane, malfunctioning machine in its unstoppable, senseless drive to jolt life into an Aryan German greatness that had never existed, and never could exist, because its blinkered puritanism was the very antithesis of all that makes a society great. Nazism could no more be great than a strutting actor in a gilt cardboard crown could be a king.

The train, huffing its way past fields of harvest stubble and woods turning golden, began to lose speed. Slowing to a crawl, it turned south and heaved into the station in the small town of Oświęcim.[21]

Shedding billows of steam, the locomotive dragged its cattle wagons up to the loading ramp. And there it stayed. Inside, the men of Buchenwald wondered if they had reached their destination yet. The hours ticked by but nothing happened. The cracks of light faded, leaving them in total darkness.

Gustav was thankful for the consolation of having Fritz by his side through these hours. How he would have coped if the boy hadn't come of his own free will did not bear thinking about. The spirit of that crushed promise of long ago lived on in Fritz, in the bond that held father and son together and had kept them alive so far. If they were indeed going to die here, at least it would not be alone.

Eventually they heard movement outside: wagon doors crashing open and barked orders; their door squealed aside and a blaze of torches and electric lanterns dazzled their eyes. "Everyone out!"

They disembarked, stiff and in pain, into a ring of light and growling guard dogs. "Form ranks! Front rank here. Quickly!" Well trained by years of roll calls, the Buchenwalders rapidly formed up in the spaces between the tracks. Expecting the usual abuse and beatings, they were surprised—and a little unsettled—to receive neither. The armed guards called out an order from time to time, but otherwise were eerily silent, walking up and down the rows, observing the new prisoners closely. Time passed, and the men grew more and more nervous. Whenever no guards were nearby, Gustav reached out and hugged Fritz to him.

The last time Gustav had set foot in this station had been in 1915, when he was discharged from the hospital and sent back to the front line. Nothing about it seemed familiar.

It was a little after 10:00 p.m. when the tramp of boots marching along the loading ramp heralded the arrival of an SS squad from the camp. They were led by a hard-faced officer, middle-aged, with a grim slant to his mouth and steel-rimmed spectacles. This was SS-Lieutenant Heinrich Josten of the Auschwitz detention department.[22] He meticulously checked off the names and numbers of the new arrivals on a list, then raised his voice: "Does any man have any watches or other valuables? Gold, for example? If so, you are to give them up. You will not need them now." Nobody responded. Josten gave the nod to his men, and they began marching the prisoners in orderly fashion along the ramp.

From the freight yard, they marched down a long, straight street between what looked like light industrial buildings and rows of dilapidated wooden barracks. Now, this *did* look vaguely familiar to Gustav.

Turning left, they went along a short road leading to a gateway flooded with arc lights; the gates swung wide, the barrier lifted, and the Buchenwalders marched in under the wrought-iron arch with its slogan:

ARBEIT MACHT FREI

Work brings freedom. The barrier descended and the gates clanged shut behind them.[23]

They were now inside Auschwitz concentration camp. They passed along a wide street flanked by trim grass verges and large, well-built two-story barrack blocks; they were similar to the SS barracks at Buchenwald, but to Gustav's eye there was a different kind of familiarity, more distant. He had been here before.

Arriving at a building in the far corner of the camp, the Buchenwalders were ordered inside. They found themselves in a bath block. Their names were checked off against the transport list again, and they went through into a changing room staffed by prisoners. Here they were ordered to strip naked for a medical inspection; following that, they would be showered and their uniforms deloused before going to their accommodation.[24]

Fritz and his father glanced at each other; the nervousness that had been growing among the Buchenwalders increased. They had heard the rumors of gassings at Auschwitz, and that the gas chamber was disguised as a shower room.[25] The men stripped off their old, soiled uniforms and underwear, then filed through yet another room, where they were scrutinized by a doctor, and another where their heads were freshly shaved— right down to the scalp, without leaving the furze of stubble they normally wore. Their bodies were also shaved, including their pubic hair. There followed a louse inspection. Fritz noticed a sign painted in sinister Teutonic letters on the white wall: "One louse is your death."[26]

Next came the showers. Fritz, Gustav, and the others watched anxiously as the first batch were herded through the door.

Minutes passed; a restlessness began to spread among the prisoners. Fritz could feel the tension mounting, marked by a low murmuring. When their turn came, would they obey and walk meekly into the lethal chamber?

Suddenly, a man's face appeared in the doorway, gleaming wet, with water dripping from his chin, grinning. "It's all right," he said. "It really is a shower!"

The next batches went through in much better spirits. Finally, they were issued with their deloused and disinfected uniforms and fresh underwear.[27] To Gustav's relief, his diary, with its pages of priceless testimony, was still secreted inside his clothes.

When they were dressed, they were inspected by SS-Captain Hans Aumeier, deputy commandant and head of Department III—the "protective custody" section, which covered most Jews. Drunk and in a foul temper, he slapped the block senior— a German wearing a green triangle—who'd turned up late to collect the new arrivals. Aumeier was everything that caused the SS to be feared: a glowering martinet with a tight little slit for a mouth and a reputation for torture and mass shootings. Once he was satisfied with the new prisoners, he ordered the block senior to take them to their accommodation.

They were placed in block 16A, in the middle of the camp. As soon as they were inside, the block senior demanded that they all hand over any contraband articles, and told his room orderlies— all young Polish prisoners—to search them. The belongings taken ranged from paper and pencils to cigarette holders and pocket knives, as well as money and sweaters—all precious items. Some of the bolder spirits—including Gustl Herzog—argued with the Poles, refusing to hand over their possessions, and were beaten with rubber hoses. Any man who spoke up got a beating. Many lost objects they had treasured—keepsakes that had kept their spirits alive, or in the case of warm clothing, had kept body and soul together through the previous winter.

At last the orderlies took the men to the bunk rooms and assigned them their places—two men to a bed, one blanket each. Gustav managed to get himself and Fritz assigned to the same bed. It was like their first night in the tent in Buchenwald; at least here there was a floor and a sound roof over their heads. But there was also the certainty that life in Auschwitz would be both cruel and brief.

אבא

On the third day they received their tattoos. This practice was unique to Auschwitz, introduced the previous autumn. They queued at the registration office; each man rolled up his left sleeve, and the tattoo was laid on his arm with a needle.

Gustav's forearm still bore the scar of his bullet wound from January 1915. The number 68523 was pricked into his skin beside it in blue ink.[28] He was entered as *Schutz Jude*—Jewish "protective custody"—his place and date of birth were set down, and his trade.[29] Having volunteered for the transport, Fritz was near the end of the list, and received the number 68629. His trade was written down as builder's mate.

Then they went back to their block. Days passed, but the Buchenwalders weren't assigned to any labor detail, and were left more or less alone, except for regular camp rituals.

There was no square, and roll call took place in the street outside the block. Food was doled out by the Polish room orderlies and the block senior—the *blockowi* as the Poles called him. The Poles hated and despised the Austrian and German Jews—both as Germans and as Jews—and made it plain that they stood no chance of surviving long in Auschwitz; they'd been sent here only to be killed. At mealtimes the Jews were made to queue up, and when a man's turn came, he was given a bowl and spoon by the *blockowi* and shoved forward. An orderly doled out a splat of thin stew from a bucket, while a young Pole stood by with a spoon, quickly scooping out any pieces of meat he spotted in the bowl. Even the most phlegmatic among the Buchenwalders were aggravated by this ritual, but any man who complained received a beating.

Gustav, who was officially regarded as Polish by birth and spoke the language, was a little better treated than others. During those first few days he became acquainted with some of the older Poles, and they told him about the ways of Auschwitz, confirming what Gustav had heard about the terrible, fatal purpose of this place.

The enclosure was much smaller than Buchenwald, with only three rows of seven blocks. This, he learned, was the main camp, Auschwitz I.[30] About a mile away, on the far side of the railway, a second camp, Auschwitz II, had been built at the village of Brzezinska, which the Germans called Birkenau—"the Birch Woods" (the SS liked picturesque names for their places of suffering).[31] Birkenau was vast, built to contain over a hundred thousand people and equipped to murder them on an industrial scale. Auschwitz I had its own killing facility: the infamous block 11—the Death Block—in whose basement the first experiments with poison gas had been carried out. Most notoriously, the enclosed yard outside block 11 was the location of the "Black Wall" against which condemned prisoners were shot.[32] Whether the Buchenwalders would be sent to Birkenau or die here was yet to be discovered.

By daylight, the familiarity of the surroundings became clearer to Gustav—specifically the well-made brick buildings. Auschwitz I had not been built by the SS; rather it had been converted from an old military barracks built by the Austrian army before the First World War. The Polish army had used it after 1918, and now the SS had turned it into a concentration camp. They'd put up additional barrack blocks and surrounded it with an electrified fence, but it was still recognizably the same place. It was here that the wounded Corporal Gustav Kleinmann had been in the hospital in 1915, in this very spot by the Soła, the river that flowed from the lake by the village where he'd been born. When he'd last seen it, it had been under snow and filled with Austrian soldiers, and he'd been a wounded hero. Treated for a bullet wound that now had a prisoner tattoo beside it.

It was as if this part of the world would not let him go; having birthed him, raised him, and nearly killed him once, it was determined to drag him back.

בן

On the ninth day after the Buchenwalders' arrival in Auschwitz, there was a demonstration of the camp's infamous character. Two hundred and eighty Polish prisoners were taken to the Death Block for execution; realizing what was intended for them, some of them fought back. They were unarmed and weak, and the SS quickly butchered the resisters and led the rest to the Black Wall. One of the doomed men passed a note for his family to a member of the Sonderkommando, but it was discovered by the SS and destroyed.[33]

"Many scary things here," Gustav wrote. "It takes good nerves to withstand it."

There were some whose nerves were beginning to fail them; one was Fritz. A sense of dread, exacerbated by the limbo in which they were being held, had been growing in him. He'd become so accustomed to his daily work as a builder, and to the fact that he owed his survival to his position in the construction detail, that being unemployed played on his nerves. He felt that sooner rather than later he would be selected as a useless eater and sent to the Black Wall or the gas chambers, as would they all. Misgiving turned to anxiety and dread. He became convinced that the only way to save his life was to identify himself to someone with authority and ask to be assigned work.

He confessed his thoughts to his father and his close friends. They argued strenuously against this rash idea, reminding him of the fundamental rule of survival that you *never* drew attention to yourself in the slightest way. But Fritz was young and headstrong, and had convinced himself that he was doomed if he did not.

The first person he approached was the SS Blockführer. With the courage of desperation, Fritz identified himself. "I'm a skilled builder," he said. "I would like to be assigned work." The man stared at him in disbelief, glanced at the star on his uniform, and scoffed. "Who ever heard of a Jewish builder?" Fritz swore it was true, and the Blockführer—unusually easygoing for an SS guard—took him to the Rapportführer, the genial-seeming Sergeant Gerhard Palitzsch.

Palitzsch was one of the few SS men who lived up to the Aryan ideal of athletic, chiseled handsomeness, and was pleasant and serene in his manner. This was a dangerous illusion. Palitzsch's record as a murderer was second to none. The number of prisoners Palitzsch had personally shot at the Black Wall was beyond counting; his preferred weapon was an infantry rifle, and he would shoot his victims in the back of the neck with an insouciance that impressed his fellow SS men. Auschwitz's commandant, Rudolf Höss, often watched Palitzsch's executions, and "never noticed the slightest stirring of an emotion in him"; he killed "nonchalantly, with an even temper and a straight face, and without any haste."[34] If any delay occurred, he would put down his rifle and whistle cheerfully to himself or chat with his comrades until it was time to resume. He was proud of his work, and felt not the slightest brush of conscience. The prisoners considered him "the biggest bastard in Auschwitz."[35]

And this was the man to whom Fritz had chosen to make himself conspicuous. Palitzsch's reaction was the same as the Blockführer's—he had never heard of a Jewish builder. But he was intrigued. "I will put it to the test," he said, adding: "If you're trying to fool me, you'll be shot at once." He ordered the Blockführer to take the prisoner away and make him build something.

Fritz was escorted to a nearby construction site. The bemused kapo provided materials and, thinking he'd fox this uppity Jew, instructed Fritz to try to make a pier—the upright section between two windows—an impossible task for anyone not properly skilled in bricklaying.

Despite the threat hanging over him, Fritz felt absolutely calm for the first time in weeks. Taking a trowel and a brick, he set to work. His hands moving quickly and deftly, he scooped mortar from the bucket and slapped it onto the first course, snaked the tip of the trowel through it, spreading the gray sludge, slicing the excess from the edges with quick strokes. He picked up a brick, buttered it and laid it, swiped off the mortar, then laid another and another. He worked with the silent speed he had learned under the gaze of SS supervisors, and the courses soon stacked up, straight, level, and even. To the kapo's astonishment, he soon had the basis of a neat, perfectly sound pier.

Within two hours he was back at the camp gate, escorted by a very surprised Blockführer. "He really can build," the man told Palitzsch.

Palitzsch's usually impassive face registered displeasure; the idea of a Jew being a builder—an honest working man—went against his sense of what was true and proper. Nevertheless, he noted down Fritz's number and sent him back to his block.

Nothing changed immediately, but then, on October 30, the eleventh day since their arrival, the moment of reckoning finally came for the Buchenwalders.

Following morning roll call, all the newly transferred Jewish prisoners were paraded for inspection by a group of SS officers. In addition to the four hundred from Buchenwald, there were over a thousand from Dachau, Natzweiler, Mauthausen, Flossenbürg, and Sachsenhausen, as well as 186 women from Ravensbrück—in all, 1,674 people.[36] They were ordered to strip naked and walk slowly past the officers so that they could be evaluated. Those who appeared old or sick were directed to go to the left, the others to the right. Everyone knew full well what being sent to the left would entail. The rate of selection appeared to be about half and half.

Fritz's turn came. As he approached, the officer in charge looked him up and down, and immediately indicated the right.

Then Fritz stood and watched as the depressing spectacle progressed. Eventually his papa's turn came. Gustav was over fifty years old and had suffered badly that year. Several hundred other men of his age—some younger—had already been sent to the left. Fritz watched with his heart thumping and breath halting as the officers looked his father up and down carefully. The hand went up—and pointed to the right. Gustav walked over and stood beside Fritz.

By the end, more than six hundred people—including around a hundred Buchenwalders and virtually all the men from Dachau—had been condemned as unfit. Many were old friends and acquaintances of Gustav's and Fritz's. They were marched away to Birkenau and never seen again.[37]

"So this was the beginning in Auschwitz for us Buchenwalders,"

Fritz would recall later. "We knew now that we were doomed to death."[38]

But not yet. Following the selection, the remaining eight hundred men were also marched out of the camp. But instead of heading west toward the railway and Birkenau, they were led east. The SS had work for them; there was a new camp to be built. They crossed the river, passing the town of Oświęcim, and marched on into the countryside.

As they marched, driven in the familiar violent fashion, the Buchenwalders felt relief out of all proportion with their circumstances. They were alive, and that was everything. Whether Fritz's intervention had precipitated this move, by planting the idea that Jews could build, nobody knew, but Gustav believed it was so. "Fritzl came with me willingly," he wrote in his diary. "He is a loyal companion, always at my side, taking care of everything; everyone admires the boy, and he is a true comrade to all of them." In at least some of their minds, Fritz's rash action had saved them all from the gas chamber.[39]

12. Auschwitz-Monowitz

✡

If an airplane were to fly east over southern Poland on a day in November 1942, the people aboard would see little trace of the German occupation. Just small country towns and old mercantile cities straddling winding roads and rivers.

Toward Cracow, a shape emerges from the fields near the brown line of the railway—a vast rectangle nearly a mile long and half a mile wide, filled with row after row after row of oblong barracks. Watchtowers dot the perimeter fence, and on the near edge, among some trees, several buildings are set apart, pouring out smoke.

Farther on, a dense cluster of buildings on the other side of the railway—the Auschwitz camp, distinguishable among the gray mass of workshops by the terra-cotta roofs of its barrack blocks. The river winds southward, a silver line fringed with deep green woods, toward the old garrison town of Kenty—where Gustav Kleinmann was stationed before the Great War—and the Beskid Mountains. Beyond, just out of sight, the lake and the village of Zabłocie, where Gustav was once a boy.

Several miles beyond Oświęcim, a new scar appears on the landscape: a vast, dark blight in a bend of the Vistula. Once, there was just the sleepy hamlet of Dwory here; now an area two miles long and almost a mile wide lies stripped, gridded with roads and tracks and filled from end to end with construction sites, dotted with offices, workshops, factory buildings, and the shells of many more, half built, laced with cages of pipework, silos, and shining steel chimneys. This is the Buna Werke chemical works, under construction and already way behind schedule.

Tucked in beside it at the far end, where the little village of Monowitz stood until the SS emptied it, lie the beginnings of a new camp. A simple oblong marked out among the fields—minuscule beside the spread of the factory complex, with just a handful of barrack blocks, a few incomplete roads and building sites, speckled with the dots of prisoners hard at work.

בֵּן

Fritz kept his mind focused on the task before him, as if it were all that existed, as if his whole world consisted of this wall, and his whole being nothing but a machine making it slowly higher and longer. The only way to stay sane was to concentrate on the minuscule, the achievable, and one's capacity to make it real.

"Tempo, tempo! Faster, faster!" The voice of the Polish kapo, Petrek Boplinsky, brayed across the site. The man knew only a few words of German, and the only one they ever seemed to hear was *schneller!* as he strode about with his cane, whipping the brick and mortar carriers. The drive to build the camp was furious; the pressure for speed came from the very top, and only the toughest and fittest could survive the pace. Few of the half-starved prisoners were up to it.

"*Pięc na dupę!*"* Boplinsky yelled, followed by the sound of the cane lashing some poor carrier five times across the buttocks. Without looking up, every other man put on a little extra speed.

A couple of weeks had passed since Fritz and the others had arrived in the Monowitz sub camp.[1] It had been a living hell, as

* "Five on the arse!" (Polish).

bad as the worst of Buchenwald. Many had not survived the initial onslaught.

After the three-hour march from Auschwitz I, the new men had been herded into their blocks. There was almost no camp—just flat, open fields with a few wooden barracks, no fence, and only a sentry line to keep the prisoners in.[2] The barracks were primitive and incomplete, with no lights or washing facilities. The only water supply was a few standpipes out in the field. There were no kitchens yet, so food was delivered each day from Auschwitz I.

At first the new men had been put to digging roads. Fritz too. The Monowitz overseers didn't seem to be aware of his skills. It rained heavily, turning the ground to mud, which was hell to dig and bogged down the wheelbarrows. The men would return to the barracks each evening soaked to the skin and exhausted. There was no heating, but the SS Blockführers and Rapportführer still expected them to report each morning at roll call with clean, dry clothes and shoes. During those first days, Fritz regarded his older and less fit comrades with concern—especially his papa. They wouldn't be able to stand this for long.

As they dug the wet mud, Fritz watched the camp beginning to take shape, with fences and the foundations for guard towers being laid; salvation, he knew, lay in getting transferred to the construction detail.

One day, SS-Sergeant Richard Stolten, Monowitz's labor manager, happened to pass near. The SS here were unusually bad tempered; there were no guard barracks yet, and they were trucked in from Auschwitz I each day in shifts; they hated doing duty at Monowitz and were easily riled. Fritz reckoned it was worth the risk; his papa would die if this went on.

Laying down his shovel he hurried after Stolten, calling out to him. "Number 68629. I'm a bricklayer," he said, speaking quickly before the sergeant could react. He indicated his work-mates. "We're from Buchenwald; many of us are skilled construction workers."

Stolten studied him, then called the kapo over. "Find out which of these Jews are builders," he said, "and take their numbers."

It had been as simple as that. At any other time, Fritz would have earned himself a beating, but the situation here was desperate. There was colossal pressure from Himmler and Goering to complete the Buna Werke and bring its factories online, which couldn't be achieved until the camp was complete. Fritz could sense the urgency.

Many of Fritz's comrades claimed to be builders in order to transfer with him—including his papa. Woodworking was among Gustav's upholstery skills, and he passed himself off as a carpenter. While Fritz laid foundations and floors, his father helped with the prefabricated sections from which the barrack blocks were constructed.

On the other side of the Oświęcim-Monowitz road, the hulking Buna Werke loomed, half built. The works belonged to the chemical giant IG Farben and when completed were to produce synthetic fuel, rubber, and other chemical products for the German war effort.[3] The war was proving far more intense and difficult than had been expected, and the demand for fuel and rubber was frantic. The company's deal with the SS gave them an unlimited supply of slave labor from Auschwitz for construction and factory work, for which they paid the SS three to four marks a day per person (which went straight into the SS coffers). Besides being cheaper than paying civilian wages, the arrangement gave the company big savings on worker facilities, sickness benefits, recreation, and other labor costs. Productivity would be lower because of the poor physical condition of the maltreated prisoners, but the company considered the savings worth it.[4] Any workers who were too sick or broken down to work could simply be sent to the gas chambers at Birkenau and replaced from the fresh intakes constantly arriving from all over Germany's conquered territories.

These intakes—many of them Jews brought directly from western Europe and Poland—hadn't been through the winnowing process of the camps and weren't as tough as the veteran prisoners. They also lacked the essential survival skills. They were rapidly broken by the pace of the labor, the abuse, the starvation, and the lack of care for the sick. By Gustav's reckoning, between 80 and 150 of these poor wretches disappeared from Monowitz each day, sent to the gas chambers without anyone ever learning their names or stories.

The transports brought heartbreaking news to Fritz; among the new arrivals were two old friends from Buchenwald: Jule Meixner and Joschi Szende, who had been transferred temporarily to Natzweiler a few months earlier. From them, he learned that Leo Moses had been murdered there. After surviving eight years in the camps, the SS had finally finished him off. The tragic injustice of it was agonizing. Fritz recalled that first encounter in the stone quarry, when Leo had offered him the little black pills, and the influence Leo had used to have him moved to the safety of Siewert's team. Poor Leo, the hardbitten, gentle-hearted old communist, had been the dearest of friends, and Fritz grieved for him.

If Fritz had learned one thing from Leo, it was that kindness could be found in unexpected places. So it proved here. The SS brought in paid workers from Germany, and for the first time since entering the concentration camps, Fritz and Gustav worked alongside civilians. These men were wary of the SS and forbidden to speak to the prisoners, but gradually they became a little more communicative. Fritz learned that they weren't dedicated Nazis, but neither were they hostile to the Nazi cause. When he tried to probe deeper into what they thought of the brutal slave driving of the prisoners, they clammed up. And yet some of them at least were sympathetic; their manner became a little warmer, and they began to leave pieces of bread lying around after lunch, and the cigarette butts they discarded were longer than before, with a good deal of smoking left in them. The civilian foreman, nicknamed Frankenstein because of his angular skull and constant ferocious expression, proved gentler than he looked; he never yelled or berated the prisoners, and his manner influenced kapo Boplinsky, who became more approachable and used his cane less on the carriers.

Gustav had a short reprieve from outdoor labor once the first few barrack blocks had been completed. Trucks arrived loaded with bunks and bales of straw. Gustav and a few others were set to stuffing jute sacks to make mattresses. He rather enjoyed himself, stitching up the mattresses faster and more neatly than any of the others.

The reprieve was soon over, and before long he was back outdoors. With the barrack walls in their part of the camp complete, he was faced with the prospect of hard labor again. Even worse was the possibility of being assigned to the Buna Werke building sites. Men who worked there returned each evening half dead, telling dreadful stories. It was like the Buchenwald quarry all over again. Often the prisoners came back on stretchers. Any man who couldn't maintain the pace was sent to Birkenau.

With calm determination, Gustav set about avoiding this fate. Each morning, when Sergeant Stolten called out the day's requirements for skilled workers, Gustav stepped forward. Whether the demand was for roofers or glaziers or carpenters, Gustav was there, swearing that he had that skill. And he managed to pull it off, day after day, bluffing his way through all kinds of building work. Fritz worried about the consequences if the SS found him out. His father shrugged it off; he was smart, and good with his hands; he believed there was no craft he couldn't master sufficiently well to evade notice by the dolts in the SS.

As more barracks were completed, they were filled with newly arrived transports of prisoners, who were sent to work on the factory sites. Conditions in the camp were horrible beyond imagining, even for veterans: overcrowded, freezing cold, and dirty. The sanitary facilities were insufficient, and dysentery began to spread. Prisoners died in frightening numbers every day.

And yet it was mild compared with what was occurring at Birkenau. Three or four transports came to Monowitz each day filled with Jews who had survived the Birkenau selections. They told awful stories about the plundering of victims by the SS: "In Birkenau they are sleeping on dollar bills and pound notes," Gustav wrote angrily, "which the Dutch and others bring with them. The SS are millionaires, and every one of them abuses the Jewish girls. The attractive ones are allowed to live; the others go down the drain."

The Polish winter set in fiercely, freezing the ground. There was still no functioning heating in Monowitz, and the cooking facilities were ramshackle. At Christmas, the ovens broke down, and the prisoners starved for two days. They didn't even get the usual crusts from the civilian workers, who were on vacation. Eventually, food had to be trucked in from the kitchens at Auschwitz I.

To their dismay, Fritz and his papa were moved to separate blocks. They met in the evenings and talked about their situation. To Fritz, it seemed that things had never been this bad before. He was losing hope. After only two and a half months in Auschwitz-Monowitz, most of their comrades from Buchenwald were dead. The Austrian *Prominenten* had all been murdered:

Fritz Löhner-Beda, lyricist of the "Buchenwald Song," beaten to death in December for not working hard enough; Robert Danneberg, the Social Democratic politician, the same fate; the lawyer and author Dr. Heinrich Steinitz . . . the list went on, all dead. The worst blow to Fritz was Willi Kurz, the boxer, the kapo from the Buchenwald gardens who had helped Fritz and his friends survive their ordeal there.

Fritz poured out all his fears to his papa when they met in the evenings. Gustav told him not to give up hope. "Hold your head up high," he said. "Lad, the Nazi murderers will not beat us!"

But Fritz was not reassured; his friends had all lived by the same courageous philosophy, and most of them were dead.

In the privacy of his own thoughts, Gustav struggled to live by his own motto. He confided his fears secretly to his diary. "Every day the departures. Sometimes it is heartbreaking, but I tell myself, *Keep your head high; the day will come when you are free. You have good friends by your side. So don't worry—there are bound to be setbacks.*" But how many setbacks could a man take? How long could he go on holding up his head and avoiding death?

Even the fittest had little chance. The Final Solution was being enacted, and even those Jews who were strong, useful laborers were being deliberately, methodically worked to death. Their labor value was of little consequence; if one died, well, that was one less Jew to trouble the world. There were a dozen more to do his work. If a person was to survive, it must be by skill, companionship, and an extraordinary portion of luck.

For Gustav, his skills and his luck came together just in time. In January he was appointed camp saddler, with responsibility for all saddlery and upholstery work in Monowitz—mostly repairs for the SS. It was indoor work, out of the savage weather, and once the heating system became operational, he was even warm.

This felt almost like safety. Gustav was acutely conscious that others were not so fortunate, and that safety never lasted long.

13. The End of Gustav Kleinmann, Jew

בן

The buildings rose in the Monowitz camp. The double electrified fence was up, the barrack blocks all but complete, and the SS barracks were under way. Through the early weeks of 1943, Fritz helped build the headquarters garage and a command post for the SS Blockführers by the main gate.

He worked alongside a civilian bricklayer. Like many of the others, this man didn't speak to the prisoners, but whereas they avoided conversation, this man wouldn't even acknowledge Fritz's existence. Day in, day out, he never said a word. Fritz grew accustomed to his eerily silent presence until one day, out of the blue, the man murmured without looking up, "I was in the moors at Esterwegen."

It was almost inaudible, but made Fritz jump. The man carried on working without missing a beat, as if he hadn't spoken.

That evening Fritz told his father and friends about this cryptic pronouncement. They understood immediately. Esterwegen had been one of the Nazis' earliest concentration camps, part of a group established in the sparsely populated moors of northwest Germany in 1933. The camps had been set up to incarcerate political enemies—mostly members of the Socialist Party. They were run by the SA, who were so chaotically brutal that when the SS took over in 1934, their behavior seemed civilized by comparison.[1] Many of the prisoners were later released, and Fritz's silent workmate must have been one of them. No wonder he was so reluctant to be sociable—he must be in constant fear of being singled out and incarcerated again.

In confiding to Fritz, the man had broken the spell. He never spoke again, but each morning Fritz would find little gifts beside his mortar tub. A piece of bread and a few cigarettes; small things, but heartwarming and potentially lifesaving.

Working alongside free civilians, receiving acts of charity, enjoying the privileged existence of a skilled worker who didn't have to slave on the Buna sites, Fritz began to recover his spirits and grow more relaxed about life. After more than three years in the camps he should have known better.

One day he was at work on the scaffolding around the shell of the half-finished Blockführers' building. He was musing on a remark his grandfather had once made; old Markus Rottenstein had been a bank clerk specializing in shorthand with the prestigious Boden-Credit of Vienna, bankers to the imperial family.[2] He had firm views on his people's status in society, believing that Jews should be elevated and civilized, and shouldn't work in manual trades. Just then, a friend of Fritz who worked on the haulage column arrived with a load of building materials and called up to him, "Hey, Fritz, what's new?"

"Nothing," Fritz replied, and indicated his surroundings. "My grandfather always used to say, 'A Jew belongs in the coffee house, not on a builder's scaffolding.'"

His laughter died in his throat when a furious German voice called from below, "Jew! Down from that scaffolding!"

Heart racing, Fritz hurried down the ladder and found himself facing SS-Lieutenant Vinzenz Schöttl, director of the Monowitz camp.

Schöttl was an unpleasant-looking brute with snake eyes in a face like dough. His main interest was in acquiring booze and luxuries for himself through the black market, yet he had a capricious, volatile nature, and when angry was absolutely terrifying.[3] Once, when some inmates were found to have lice, Schöttl had had the whole block—including the seniors—sent to the gas chambers. He glared at Fritz. "What were you laughing at, Jew?"

Standing to attention and whipping off his cap, Fritz replied, "Just something my grandfather said."

"What did your grandfather say that was so funny?"

"He said, 'A Jew belongs in the coffee house, not on a builder's scaffolding.'"

Schöttl stared at him. Fritz hardly dared to breathe. Suddenly the dough face split and let loose a guffaw. "Clear off, Jew-pig!" said Schöttl, and walked off, laughing.

Sweating, Fritz climbed back up the ladder. He had nearly paid the price for complacency. There was no such thing as safety.

בן

The influx of Jews to Monowitz kept growing. Fritz and the other veterans were troubled by how naive some of them were. They had been through the selection at Birkenau, and their wives, mothers, children, and fathers had all been sent one way, while they—the young men—had been sent the other. They had no inkling of what would happen to their families, and hoped they would see them again.

Fritz couldn't bear to reveal the truth and shatter their hopes. Eventually, inevitably, they found it out anyway—their wives and little ones, their mothers and sisters and fathers, had all been gassed. Some of them fell into a depressed torpor. In their hearts, they died. They moved about in a state of utter apathy, didn't look after themselves, and gradually joined the ranks of the hopeless, wasting away to skin and bone, scabbed, with lifeless eyes and empty souls. In camp slang these walking dead were known as *Muselmänner*—Muslims. The origin of the term was lost in camp lore, but some said it was because when these poor souls could no longer stand, their collapsed posture resembled a Muslim at prayer.[4] Once a person became a *Muselmann*, the other prisoners would avoid him; their hearts closed, partly in disgust, partly in dread at the thought that they too might become like this.

The building work complete, Fritz was among a lucky group of six selected by Stolten to work on the camp bath block. He cemented and mounted heating units under a civilian foreman who nearly drove him mad. Jakob Preuss was all noise and bluster in front of the SS. He yelled constantly at the prisoners, and if a guard or an officer came near, Preuss would throw out a salute and cry "Heil Hitler!" He grated on Fritz's nerves.

One day, Preuss called Fritz into his office. "What do you think you're doing with your work rate?" he demanded. Fritz was taken aback; he knew better than to slack, and his performance had never been criticized before. Preuss lowered his voice and said, "If you keep working this fast, we'll be finished soon, and I'll be sent to the front!"

Fritz didn't know what to say. He was in a bind. If the work slowed, all the prisoner workers were at risk from the SS. On the other hand, if Preuss concocted some pretext to report him in revenge, that would be fatal. Fritz decided that the safer course was to slow down. Preuss became positively friendly, wangling extra food for his workers. He was joined in this by another of the German civilians, a welder from Breslau called Erich Bukovsky. Both confessed that they hoped the Nazis would be defeated.

It was beginning to look like that might happen. Until now Germany had seemed unbeatable. And then, in February, news came through the grapevine that the German force at Stalingrad had surrendered to the Russians. The Nazis were not invincible.

Fritz heard this heartening news from a French civilian named Jean, whom most people called simply "Moustache" after his extravagantly waxed face ornament. Jean also told him stories of the French Resistance. Fritz shared the information eagerly with his father and friends when they met up in the evenings. And yet, Stalingrad, Britain, and Africa—the places where the Allies were beating the Germans—were all a very long way from Auschwitz.

אבא

Gustav's fingers worked skillfully at a leather panel, trimming, pushing the heavy needle through the tough, pliant material. He was content in his day to day existence, if not in his heart. There was no shortage of work, and he was, in effect, a kapo now, with a handful of semiskilled workers under him. Being indoors had been a boon during the winter months, and even with May beginning and summer on its way, it was infinitely better than being on the haulage column or the factories.

Taking each day as it came, Gustav reassured himself that he would survive. Fritz didn't share his father's sanguine, dogmatic principle of determined optimism; he never ceased worrying about everything—his friends, his papa, the future. He worried about Edith and Kurt and fretted about what had become of his mother and Herta. Hearing the tales out of Birkenau—especially the terrible rumors leaked by the "bearers of secrets" who served in the Sonderkommando in the crematoria—it was sickeningly easy to imagine. An anger was growing in Fritz, born out of helplessness. His nature was not like his father's. Gustav tried not to dwell on things. He kept his head down, did his work, and lived from day to day. For Fritz, it wouldn't be long before his hatred of the Nazis became too great to contain. What kind of explosion might happen then didn't bear thinking about.

His thoughts focused elsewhere, Gustav had no idea that while he sat stitching, a short distance away, across the road and railway line in the Buna Werke, a decision was brewing that threatened to bring his relatively comfortable existence to an abrupt end.

Construction of the factories was still far behind schedule,[5] and a group of officers had been sent from Berlin to investigate. Himmler wanted answers. They were given a guided tour of the site by Lieutenant Schöttl and senior staff from IG Farben. What they found didn't please the SS top brass at all. The vast complex was only half complete, and no units were ready to begin production. The methanol plant was almost set to go online, but the far more important rubber and fuel plants wouldn't be ready for months, maybe another year.

They grew more displeased by the minute. It was noted that about a third of the construction workers were camp inmates, who were visibly weaker and less efficient than the paid civilians. Their effectiveness was hampered further by the necessity of constantly guarding them and keeping them together. But what really disgusted the visitors was that many of the prisoner foremen were Jews. Schöttl explained that he didn't have enough Aryans in Monowitz; nearly all the prisoners he was sent were Jews. The visitors glowered and said it would not do; Jews must not be put in positions of responsibility. They ordered Schöttl to do something about it.

A few days later at evening roll call, Schöttl appeared in company with SS-Captain Hans Aumeier, the malignant demon who had first welcomed the Buchenwalders to Auschwitz. Schöttl's porcine face looked grave, as if he had a very serious task to perform. He mounted the podium, took out a piece of paper, and read out the numbers of seventeen prisoners, ordering them to step forward from the ranks. Among them was prisoner 68523: Gustav Kleinmann. All were Jews who held foreman positions—mostly veteran Buchenwalders and Sachsenhauseners.

Everyone guessed what this must signify; such selections happened all the time, and meant only one thing: departure for Birkenau and the gas chambers.

Aumeier inspected the selected men closely, looking with distaste at the Jewish stars on their uniforms. In most cases, they were of the two-color kind: a Star of David made up of red and yellow triangles, dating from the old days when the Nazis still required a pretext for sending Jews to the camps.

"Get rid of them," Aumeier ordered.

A kapo who was standing by unpicked Gustav's star from his jacket, separated the two triangles, and gave the red one back to him. The same was done to the other sixteen men, leaving them clutching their red triangles, utterly mystified.

"You are political prisoners," Aumeier announced. "There are *no Jews* in positions of authority here. Remember this. From this moment you men are Aryans."

And that was that. As far as the regime was concerned, Gustav Kleinmann was officially no longer Jewish. By the mere alteration of a list and a badge, he officially ceased to be an intrinsic threat and burden to the German people. And there, neatly played out in one simple, self-satirizing ritual, was the whole towering idiocy of Nazi racial ideology.

From that moment, life for the Jews in Monowitz was transformed. The seventeen Aryanized men were now on a higher plane, and although they weren't immune from punishment, they were safe from outright persecution and no longer beasts in the eyes of the SS.

With their positions as foremen and kapos secure, they were able to gain influence and help their fellow Jews acquire good positions. (With the ritual over and the top brass gone back to Berlin the blanket ban on Jews holding functionary positions was quickly forgotten by Schöttl.) Gustl Herzog became a clerk in the prisoner records office, eventually rising to be its head functionary with a staff of several dozen prisoners.[6] Jupp Hirschberg, another Buchenwalder, became kapo of the SS garage, where staff cars and other vehicles were maintained; he became privy to all manner of gossip from the chauffeurs, as well as intelligence about events in the wider camp and the world outside. Others acquired jobs ranging from block senior to carpenters' kapo to camp barber. Between them, they brought the conditions for other Jews to a new level. The new Aryans were able to speak out to prevent beatings, obtain decent rations, and resist the brutal green-triangle kapos.

For Gustav it meant his comfortable working life was given an additional security. There was little danger now of his being selected for the gas chambers, and so long as he was careful, he would be safe from random acts of violence by the SS.

But his change of status had an unforeseen effect, and it was heartbreaking. He and Fritz, living in separate blocks, had grown so used to meeting in the evenings after roll call that they thought little of it; it was routine, habitual. One evening they were so deep in conversation—reminiscing about the old days, weighing up the future, exchanging news about the camp—they failed to notice an SS Blockführer eyeing their intimate conversation with deep suspicion.

He interrupted, shoving Fritz hard. "Jew-pig, what d'you think you're doing, talking to a kapo like that?" Fritz and his papa both jumped to attention, startled out of their wits. "What d'you mean by it?"

"He's my father," said Fritz, bemused.

Without warning, the Blockführer's fist slammed sickeningly into Fritz's face. "He has a red triangle; he can't be a Jew's father."

Fritz was stunned, pain rebounding through his skull; he'd

never been punched right in the face like that. "He *is* my father," he insisted.

The Blockführer punched him again. "Liar!"

Fritz, absolutely bewildered, couldn't help repeating his answer, and received another savage blow. Gustav stood by in horror, helpless, knowing that if he intervened it would make things worse for both of them.

Fritz was knocked to the ground by the enraged Blockführer, who finally ran out of steam. "Get up, Jew." Fritz picked himself up, bruised and bleeding. "Now get the hell out of here."

As Fritz walked away, nursing his head, Gustav said to the Blockführer, "He really is my son."

The Blockführer stared at him as if he were a madman. Gustav gave up; if he'd told the man he was in fact an Aryanized Jew, it would probably change nothing. Indeed, it was quite possible that the Blockführer knew that already, but would still think the same way. The mind of a Nazi was beyond fathoming, let alone reasoning.

אחים

Auschwitz-Monowitz, now completely built, was a small, simple camp. It had no gatehouse: just a plain gateway in the double electrified fence. A single street ran the length of the enclosure, a distance of just a third of a mile.[7] Barrack blocks lined the road: three rows to the left, two to the right. About halfway along was the roll-call square, with a smiths' workshop and kitchen block to one side. There was a grass border, carefully tended, as were the verges and flower beds in all concentration camps; the contrast between the care given to these patches of decoration compared with the abuse and murder of human beings was a paradox that drove some prisoners mad.[8]

A little farther along, on the left-hand side of the street, stood block 7. Outwardly it was no different from the others: a wooden barrack, not particularly well made. But inside it was very special, for this block belonged to the Monowitz *Prominenten*. These were not like the *Prominenten* Fritz had known in Buchenwald; there were no celebrities or statesmen here; just the kapos, foremen, and men with special duties—the functionary prisoners, the inmate aristocracy.[9] Gustav Kleinmann, camp saddler and new-minted Aryan, was one of them. Having come here as the lowest of the low, he was now among the most privileged.

In his personal contentment, Gustav was slowly becoming less conscious of the sufferings of others, or at least less disturbed by them. He worked indoors, and the abuse happened mostly out of his sight. On the rare occasions he took out his diary, it was to record how peace had settled on the camp, and that fewer prisoners were being sent to the gas chambers—albeit because the selections at Birkenau were becoming more thorough in weeding out and murdering the weak. By Gustav's reckoning, about 10 to 15 percent survived from each transport—"The rest are gassed. The most gruesome scenes play out." But still, "Everything is more peaceful in Monowitz, a proper work camp." To Gustav's experienced eye, its primary purpose was to exploit, not to destroy its inmates, and the horror of life within its fences was diminished compared with what he had seen. It was as if he had finally lost the ability to perceive it all in comparison with the normal, civilized world.

Even so, two things weighed on him severely. One was separation from Fritz. The other was the man who hovered above all the *Prominenten* like a malevolent, bloodsucking bat: Josef "Jupp" Windeck, the camp senior and chief of all the kapos and functionary prisoners. The SS could not have picked an enforcer more suited to their ideal than Jupp Windeck.

He wasn't much to look at—small and slight, with the bearing of a weakling. But his appearance was belied by the temperament of a tyrant.[10] His bland, characterless features expressed disdain and scorn; he loved to lord it over his fellow men and trample them down to enhance himself. A German, Windeck had been a petty criminal since the age of sixteen, in and out of prisons and concentration camps since the early 1930s. He wore the black triangle of an "asocial," a catch-all that included addicts, alcoholics, the homeless, pimps, the unemployed, and the "immoral." A camp senior in Auschwitz I, he'd been transferred to Monowitz along with the Buchenwalders.

In no time at all he'd established a reign of corruption, terror, and extortion. "Well, so much stuff came with the Jews," Windeck recalled later, "and we filched from it, of course we did . . . as kapos we always got ourselves the best."[11] His chief ally was an SS Rapportführer called Remmele, who benefited from Windeck's moneymaking schemes.

Windeck dressed as he liked, favoring riding boots with breeches and a dark jacket—probably in an attempt to mimic the look of an SS officer. He strutted self-importantly about the camp, never without his dog whip. There were allegations that he sexually abused younger prisoners. He murdered with impunity, beating or kicking his victims to death or drowning them in the washroom basins.[12] It was Jupp Windeck who had murdered the lyricist Fritz Löhner-Beda, lashing the weakened, broken old man with his dog whip.[13] His henchman described how he "particularly liked to beat up feeble, half-starved, and sick inmates . . . When these miserable fellows lay on the ground before him, he trampled on them, on their faces, their stomachs, all over, with the heels of his boots." He was extremely proud and vain about his riding boots: "God help the man who dirtied Windeck's boots, for he could be murdered for that."[14]

Gustav and his high-status friends were able to hold off Jupp Windeck's cruelties and protect their fellow Jews. They were helped by the communist prisoners, with whom they formed an alliance.[15]

The balance of power swung against them when a transport of six hundred prisoners arrived from Mauthausen, reputedly one of the regime's harshest camps. They were all green-triangle men, with some real savages among them. Windeck quickly gathered them around him, steering them into positions as kapos and block seniors. The Aryanized Jews and the communists resisted, but Windeck and his cronies were too powerful. Any prisoner who showed fight was beaten—sometimes to death. The misery in Monowitz redoubled.

Relief only came when Windeck's barbaric green men began to fall into traps of their own making. One would go on a drunken bender, another would steal from the camp, yet another would pick a fight with an SS guard or a civilian worker. They were removed and sent to the unspeakable purgatory of Auschwitz's coal-mining subcamps.[16] As the months went by, Jupp Windeck's power base eroded until eventually it was gone.

It was Windeck's own corruption that brought on the final crisis. Gustl Herzog, in his position as clerk in the prisoner records office, discovered evidence that Windeck had acquired a precious necklace and was intending to post it to his wife. This intelligence was conveyed to the camp Gestapo at Auschwitz I. Windeck was seized and sentenced to two weeks in the bunker, after which he was sent to a punishment company in Birkenau. He never troubled Monowitz again.[17]

Gustav and his friends regained their influence. The atmosphere among the prisoners became comradely again; they received their proper allowance of food, had showers once a week and fresh laundry once a month. There was order, and all that remained to worry about were routine hazards: the SS, sickness, the ceaseless dangers of work, the periodic selections of the ill and weak for the gas chambers. By contrast with what they had just been through, it could almost be called civilization, albeit a civilization carved out with bleeding fingers inside the fences of hell.

14. Resistance and Collaboration: The Death of Fritz Kleinmann

בן

The Nazi system was a formidable but ramshackle piece of engineering. It had been built through improvisation and ran at a juddering pace, misfiring, stuttering, consuming its human fuel, pouring out bones and ashes, and ejecting an exhaust of nauseating smoke. The individual human, in drab stripes, was forced not only physically into the machine but morally and psychologically too. Beyond the Blockführers and kapos, the electrified wire and watchtowers, the SS commandants and guard dogs, beyond the roads and railways, the camp system and the hierarchy of the SS, stood a whole nation, a government and society of human beings whose base, animal emotions—fear, spite, lust for gain or some imagined former greatness—empowered the system.

The prisoners' incarceration was meant to be the clean, simple solution to the society's complex, muddy problems. The removal of human toxins—criminals, left-wing activists, Jews, homosexuals—was supposed to bring back the nation's glory days. In fact it was not a cure but a poison, slowly but surely bringing the nation to the ground. The inefficient labor of starved slaves, the cost of the system that enslaved them, the weakening of science and industry by the removal of geniuses tainted by race: all these things hamstrung the nation's economy. Becoming a pariah among nations had cost trade. Germany tried to solve these further problems by wars of conquest, more enslavement, more murder of the people believed to be the root cause, the stone crusher rattling on, day and night, grinding and destroying and slowly wearing itself out.

Fritz Kleinmann found the helplessness and hopelessness of being trapped in the machine intolerable. His father was safe for now, which lifted a great weight from his heart. But the injustice and cruelty of the system could make a sane man crazy, and a pious one curse God. They lived, and in most cases died meaningless deaths, within fences and walls built by their fellow prisoners. Fritz himself, with meticulous skill, had helped create this prison out of open fields. The very bricks and stones that Fritz laid had been molded and cut by yet other prisoners in the brickworks and stone quarries run by the SS.[1]

The bond he shared with his father, and their ties with their friends, were far from universal traits. Solidarity and cooperation, the keys to survival, rarely come naturally to men in extreme circumstances. Deprivation and hunger bred hostility between prisoners, to the point where they would fight over an unfair portion of turnip soup, where a person might commit murder for a piece of bread. Even fathers and sons had been known to kill one another in the extremity of starvation. Yet only through solidarity and kindness could people stay alive for any length of time. Lone wolves and mavericks, or those unfortunates who were isolated by their inability to understand German or Yiddish, never lasted long against the relentless terror.[2]

It took strength of character to share and love in a world where selfishness and hate were common currency. And survival was never guaranteed. Fritz saw the marks of abuse and deprivation and the signs of impending death in all his fellow prisoners, including himself:[3] bruises, cuts and broken bones, sores and scabs, pallor and chapped skin, limping gait and gapped teeth.

The prisoners were able to shower once a week, but it was an ordeal. Those with harsh block seniors had to strip in the bunk room and then run naked to the shower block. After showering, only the first men out got dry towels; they were passed along, so if you lagged behind you got nothing but a soaking rag and had to walk back to the barrack dripping, even in the coldest winter weather. Pneumonia was endemic, and often fatal. There was a prisoner hospital, but although it was kept decently equipped by its prisoner staff,[4] treatment under the SS doctors was rudimentary, and it was a fearful place, often full of typhus patients. Nobody went there unless they had to; patients were subject to selections, and if deemed unlikely to recover quickly, they went to the gas chambers or received a lethal injection.

Food was distributed in the barrack. Only a few bowls were provided, so the first to get their helping of soup had to wolf it down so as not to keep the others waiting. Any man who took his time would be shoved impatiently. Their acorn coffee was served in the same bowls. If you managed to acquire your own spoon, it was as precious to you as jewels; you would guard it with your life, and as knives were unobtainable, you would extend its usefulness by sharpening the handle on a stone. There was no toilet paper in the latrines, so scrap paper was another valuable commodity; torn-up cement bags from the construction sites could be obtained, and sometimes a newspaper might be acquired from a civilian—perhaps left lying around at the factory and smuggled back to camp. Pieces could be used or traded for food.

The people suffering this degradation were regarded by the Germans as human garbage, but the nation's war economy was increasingly dependent on their labor. This was the new age of greatness that Hitler had brought into existence: a world in which a square of waste paper became a currency with a tangible value, either to spend or to keep one's ass wiped.

Each man's body was subjected constantly to shocks and irritations. Having a decent pair of shoes was absolutely fundamental. If they were too large or too small, they chafed and caused blisters, which were prone to infection. Socks were rare, and many substituted strips of fabric torn from the tails of their camp-issue shirts. This in itself was risky, because damaging SS property was sabotage and could earn you twenty-five lashes or a period of starvation. With no scissors or clippers, toenails grew and grew until they broke or became ingrown.

Heads were shaved every two weeks by the camp barber. Partly this was to prevent lice, but it also served, like the striped uniforms, to make prisoners conspicuous. The barber used no soap or antiseptic, so every man's head and face had razor burn, pimples and pustules, as well as ingrown hairs. Infections were common, and could lead to time in the hospital. Fritz was at least spared half the shaving ordeal—at twenty years old, his beard had still not developed.

There was a camp dental station, but prisoners didn't go there if they could help it. Loose fillings led to cavities and gum disease, while scurvy brought on by poor diet loosened teeth. Gold teeth could be lifesavers or a deadly danger. Prisoners were murdered for them by certain kapos, but if the owner of a gold tooth possessed the strength of will to pull it out himself, it could be traded for luxuries. There was a fixed exchange rate among the civilian black marketeers: one gold tooth equaled one bottle of Wyborowa, a quality brand of Polish vodka. Or it could buy five big loaves of *Kommisbrot** and a block of margarine. These things could be traded onward for other goods. In a world where each week, each day, or even each hour might be one's last, there was little point in storing up riches for some better or higher purpose. Anything that brought solace or comfort or a full stomach in the living moment was worth the price.

For the managers and board of IG Farben, the sacrifice of their slave workers was justified by profit. Some of the staff felt guilt,

* Military-ration bread made from sourdough with a long shelf life.

but it was minimal and ineffectual. Meanwhile, the accountants and directors turned a blind eye to the huge quantities of their delousing chemical Zyklon B purchased by the SS, especially at Auschwitz, where its toxic fumes fed the gas chambers.[5]

Fritz Kleinmann was in no doubt where the evil came from: "Let no one conclude that the prisoner hierarchy bears the blame for bringing about this state of affairs. Some of the functionary prisoners adapted themselves to SS practices for their own profit, but the sole responsibility belongs to the SS killing machinery, which achieved its perfection in Auschwitz."[6] Each prisoner who passed the selection at Birkenau could expect to survive, on average, for three to four months.[7] Fritz and his father had so far lasted more than eight. Less than a quarter of their four hundred tough, seasoned comrades from Buchenwald were still alive.

Although Auschwitz had achieved a kind of industrial perfection, as a machine it was flawed, inefficient, and liable to failure. Its very brutality created in some a will to resist, and its corruption produced the cracks and flaws that allowed resistance to thrive.

During his first summer in Auschwitz-Monowitz, when Jupp Windeck's dominance was at its height, the resilience and moral indignation that were defining parts of Fritz's character led him to become involved in the resistance. In doing so he was putting his life in jeopardy. But he did that every day just by existing; every little scrape or misplaced glance or bout of freezing weather or contact with disease could start a chain reaction leading to incapacity and death. By resisting, it was at least possible to risk everything *for* something.

בן

It began with a conversation in a quiet corner of the barrack and ended in a new job.

Construction work in the camp was complete by summer 1943, and the need for builders at the Buna Werke was declining. Fritz was at risk of outliving his usefulness. Certain friends of his decided that he could both be preserved and be of use to them. They took him aside and spoke to him in utmost secrecy.

They were Buchenwalders he'd known for years. There was Stefan Heymann, Jewish intellectual, war veteran, and communist, who'd been like a second father to Fritz and the other boys. Also present was Gustl Herzog, along with Erich Eisler, an Austrian antifascist. They had a task for Fritz—a vital and potentially dangerous one.

Throughout their years in the camps, these men had been involved in a covert Jewish-communist alliance against the SS. Their resistance consisted mainly of acquiring positions of influence in order to gain information useful to their comrades' welfare and survival. It was partly through the efforts of this network that Fritz and Gustav had been moved to less dangerous work details, that Robert Siewert's builders' school had been set up, and that Fritz had learned of the contents of his mother's last letter and had advance warning that his father was listed for Auschwitz.

The resistance had reestablished itself in Monowitz, placing its members in important functionary jobs thanks to the Aryanization of friends such as Gustav. But now they felt that they ought to escalate their activities. Acts of minor sabotage were all very well; Fritz participated in such acts on the building sites—a bag of cement dropped heavily so that it burst, a running hose surreptitiously hooked over the side of a truck loaded with cement—but the organized resistance wanted to do more.

Information was the key. Functionary prisoners were able to obtain all manner of intelligence about the other Auschwitz satellite camps, prisoner movements, selections, and mass killings.[8] Now they wanted Fritz to help them open up another valuable source: civilian workers. It would involve having him transferred to one of the factory details inside the Buna Werke. He'd shown himself good at making friends with civilians, and thousands of them worked in the factories. A place was found for him in Schlosserkommando 90—the locksmith section of the construction command.

And so it happened one morning that, for the first time since arriving in Monowitz, Fritz went beyond the camp perimeter, marching with the labor force and their SS guards out of the gates, across the main road, and along the lane leading to the Buna Werke.

Only upon entering the site did he realize just how vast it was. The whole complex was a grid of streets and railway spurs. A person standing on one of the main east-west streets could scarcely make out its far end in the haze nearly two miles away. The cross streets, running north-south, were over half a mile long. The rectangular lots were packed with factory buildings, chimneys, workshops, depots, oil and chemical storage tanks, and weird structures of pipework looking like truncated sections of fairground rides. The complex was divided into sections: the synthetic oil plant with all its supporting workshops, the Buna rubber factory, the power plant and smaller subsections to manufacture and process chemicals. Most of it was still half dormant—the structures built but the internal workings far from complete.

Several thousand men and women worked in the factories. About a third were prisoners, the rest civilians. The locksmith section—which in fact undertook a variety of metalworking jobs in its workshop and around the factories—turned out to be a friendly, easygoing team. The prisoners were treated kindly by most of their kapos, and encouraged to "work with the eyes," taking it slow while keeping a sharp eye out for the slave drivers.[9] Fritz's kapo was a sympathetic political prisoner, a former Dachau man, who had helped arrange his work placement for the resistance.

Fritz was made a general assistant in a subsection on one of the main factory floors,[10] where there were a great number of German civilians—mostly engineers, technical workers, and foremen. The majority of their laborers were Polish and Russian prisoners, who found it hard to follow instructions in German and were treated abominably by their kapos. If the civilian foremen weren't satisfied with the workers' performance, IG Farben had them sent to Auschwitz I for "reeducation." German-speaking prisoners had it much easier; Fritz became known to the civilian foremen, and gained their trust.

He developed a sympathetic relationship with one in particular. Again he received discreet gifts of bread and cigarettes, or occasionally a newspaper. From time to time the German stopped by for a brief chat, and Fritz listened eagerly to his news about the war, which flatly contradicted the propaganda. Things were going badly for Germany on all fronts; having lost Stalingrad, they were being battered ferociously in the east, besides getting kicked out of North Africa by the British and Americans, who would soon be in Italy and driving north toward Germany. It was clear to Fritz that this German was no Nazi; he hoped fervently that the war would end soon, and that Germany would lose. Each day, Fritz carried back verbal reports to his comrades (along with the valuable gifts of bread and newspapers).

Although he knew his task was important as well as hazardous, Fritz had little idea of the scale of the operation he'd become involved in. From disorganized beginnings, the Auschwitz prisoner resistance had lately become an efficient, coordinated network. On May 1, 1943—a Nazi holiday when the SS operated a skeleton staff—a secret meeting had been convened in Auschwitz I, at which two resistance factions agreed to cooperate. They were dominated by a Polish group, including a number of former army officers, under the leadership of Jósef Cyrankiewicz, who persuaded his people to cooperate with the Jews and the Austro-German politicals. This combined all their various advantages—the Germans' understanding of Germany and the Nazis, which was vital in intelligence, and the fact that Polish prisoners were allowed to receive mail, which enabled them to bring in supplies and communicate with local partisans.

They called themselves Battle Group Auschwitz—a measure of their militancy[11]—and soon established contact with Stefan Heymann and the Monowitz resisters. Intercamp cooperation was facilitated by the constant shuffling of prisoners and labor details around the complex. What the Monowitz group brought to the table was its ability to cultivate relationships with civilians and disrupt production at the Buna Werke. Sabotage was extensive and constant. Prisoners in the electricians' detail had managed to short-circuit a turbine in the power plant. Another group, taking advantage of the reduced guard on May 1, had caused an explosion in the half-complete synthetic fuel plant, while others destroyed fifty vehicles.[12] These acts, together with a general go-slow, had helped delay the completion of the various factories.

Of all resistance activities, civilian contact was among the most dangerous. The camp Gestapo was constantly endeavoring to penetrate the resistance and expose its leaders and members. Thus the work of spotting and weeding out informers was unending. This was especially vital when it came to the most sensitive resistance operation: the planning and execution of escapes.

While Fritz went back and forth between factory and camp each day, carrying his little snippets of intelligence, he was only dimly aware of his connection with this network and the significance of his role in it.

בן

It was a Saturday in June, and the working day was over. At evening roll call, the prisoners stood to attention in the knowledge that tomorrow, although not exactly a day of rest, was at least a day of less toil and reduced danger.

Fritz stood in his place, uniform buttoned neatly, cap on straight and flattened to one side in the approved beret style, ready to whip it off mechanically at the "Caps off!" order. Everything was normal, the same slow, monotonous, grinding, day-in, day-out repetition he had known twice daily since October 1939, almost without variation.

The Rapportführer had completed his duties and was about to dismiss the parade when he noticed a small knot of figures entering the square, and paused. As the figures reached Fritz's field of vision, he made out two SS sergeants force-marching a man who limped and stumbled ahead of them. Fritz peered curiously, sidelong, keeping his face to the front. They shoved and hit the man as they would a prisoner, but he wasn't in uniform and his head wasn't shaved. He appeared to be a civilian, but he'd been violently worked over, his face bloodied and swollen. As they came closer, with a sickening jolt Fritz recognized his German contact from the factory. The SS men escorting him were Staff Sergeant Johann Taute, head of the Monowitz subdivision of the camp Gestapo, and his subordinate Sergeant Josef Hofer.

Fritz watched in silent, mounting horror as they forced the civilian to face the assembly and ordered him to identify any and all prisoners he'd had contact with at the factory.

The man peered at the thousands of faces before him. Fritz, buried deep in the mass, was well out of sight. With the two SS men pushing him along, the civilian went between the ranks, back and forth, studying the faces up close. He came along Fritz's row. Fritz stared straight ahead, heart thumping. The bruised, bloodshot eyes looked at him reluctantly, and a hand rose and pointed. "This one."

Fritz was seized and, together with the civilian, force-marched past his friends and comrades, past the horrified eyes of his father, and out of the square.

בן

He was bundled into the back of a truck and driven from the camp. The truck drove the few miles to Auschwitz I, but instead of entering the camp compound it pulled up in front of the Gestapo building, which stood outside the fence, opposite the SS hospital and beside a small underground gas chamber. Fritz was marched along a corridor by Sergeants Taute and Hofer and pushed into a large room.

In sick terror, Fritz took in the spartan fittings of the room. There was a table with straps attached to it, and hooks embedded in the ceiling above. He'd lived long enough in the camps to know what these things were for.

After a while an SS officer entered the room. He looked at Fritz with lively, smiling eyes in a gentle, patrician face. Prematurely bald and graying, SS-Lieutenant Maximilian Grabner didn't look at all threatening; indeed, he looked like a university professor or a genial clergyman. Rarely could a man's appearance have been more at odds with his character; the affable-looking Grabner was head of the Auschwitz Gestapo, and his reputation for coldly, pitilessly instigating mass murder was unsurpassed in this or any other camp. He regularly purged the hospital and the camp bunker—"dusting off" he called it—sending the inmates to the gas chambers or the Black Wall. He'd instituted a program of exterminating pregnant Polish women, and was reckoned personally responsible for over two thousand murders. There were few men in Auschwitz as feared as Maximilian Grabner.[13] He terrified even the SS.

He studied Fritz a moment; then he spoke. His voice was eerily soft, and his accent was redolent of the rural areas outside Vienna, simple and uneducated.[14]

"I know," he said matter-of-factly, "that you, prisoner 68629, are involved in planning a large-scale escape from the Auschwitz-Monowitz camp, and that you've been doing so with the collaboration of the German civilian who pointed you out. Sergeant Taute's men have been keeping an eye on him. His irregular behavior caught your attention, didn't it, sergeant?"

Taute nodded, and Grabner turned his friendly gaze back to Fritz. "What d'you have to say about that?"

Fritz had no idea what to say. He couldn't deny knowing the civilian, but the stuff about an escape was a total mystery.

Grabner took out a notepad and pencil. "You will now give me the names of all the prisoners involved in this plot."

Taking Fritz's stunned silence for a refusal, Grabner nodded to Taute and Hofer.

The first blow of Hofer's cudgel bent Fritz double and knocked the breath out of him; then came a second and a third.

But no confession was forthcoming. Grabner was surprised. Although little more than a boy, it seemed that prisoner 68629 would be harder to break than the civilian had been. At a gesture from Grabner, the sergeants pushed Fritz facedown on the table and fastened the straps over his body, pinning him. The cane rose and flashed down, humming, lashing him across the buttocks. And again, and again, until his backside was lacerated and on fire with agony. Even in this extremity of fear and pain he kept count of the lashes: twenty searing blows before they unstrapped him and stood him up.

"Admit what you've done," Grabner said, indicating the notepad. "Give me the identities of the prisoners you were planning to help escape."

Fritz knew that denying it would be useless, and so he said nothing. Again he was forced down on the table, again the straps were fastened, again the cane hummed in the air.

He lost track of how many times he was strapped down, but he doggedly kept count of the blows: altogether, sixty searing weals on his flesh.

They unstrapped him and hauled him to his feet again. He could hardly hold himself upright. Grabner studied him closely. "Tell me the names."

Sooner or later, the point would come—as it would to any human being trapped in this nightmare—when Fritz would crack and say whatever it took to make it stop. Truth or lies—it wouldn't matter, so long as the torture ended. He could name his friends who were involved in the resistance. It would be simple, and he wouldn't be human if he didn't feel the temptation. Stefan, Gustl, Jupp Rausch, and the other resisters, his friends and mentors: he could condemn them to torture and death. Fritz retained sense enough to know that it wouldn't really save his life, but it would at least halt the torment.

He said nothing. Grabner nodded at Taute and Hofer, and indicated the hooks in the ceiling.

Fritz's wrists were yanked behind his back and tied so tightly that the circulation was cut off. The long end of the rope was thrown up over a hook, and the two sergeants hauled on it. Fritz's arms were wrenched backward and upward, and with an indescribable, blinding agony he was lifted off his feet. He hung with his toes a foot above the floor, his body weight wrenching his shoulders in their sockets, filling his mind with screaming pain. He had seen many poor souls suspended like this from the Goethe Oak, but the experience of it was worse than could ever be imagined.

"Give me the names," Grabner repeated, again and again. Fritz hung for nearly an hour, but all that came out of his mouth was incoherent squeaks and drool. "You won't live through this," Grabner's voice said softly in his ear. "Give up the names."

At a nod from Grabner, the rope was let go, and Fritz crashed to the floor. Grabner repeated his question over and over: name names and it would be finished. Still Fritz said nothing. They dragged him to his feet, hauled again on the rope, and raised him screaming into the air.

Three times they hung him, without result. Grabner was losing patience. It was Saturday night, and he was keen to get home. This interrogation was wasting his precious leisure time. Fritz had hung for an hour and a half all told when they let go and he crashed to the floor a third time. He was dimly aware of Grabner leaving the room, ordering the two sergeants to take the prisoner back to the camp. The interrogation would resume later.[15]

בן

After Fritz had been taken away, Stefan Heymann and the other resisters fretted and talked over what they should do. How long did they have before Fritz was broken and the Gestapo came back for the rest of them? All that evening they debated, trying to plan for the catastrophe heading their way.

Gustl Herzog was still up and about when he heard that Fritz was back in camp. He rushed to meet him, and found him being carried along the street by two old Buchenwald friends: Fredl Lustig, an old comrade of Gustav's, from the haulage column, and Max Matzner, a near-victim of the infamous typhus experiments.

Fritz couldn't stand; aside from the visible bruises and blood, his joints and back were in excruciating pain. Gustl told Lustig and Matzner to take him to the hospital, then went in search of the other resisters.

The hospital occupied a group barrack block in the northeast corner of the camp. It had several departments: medical, surgical, infectious diseases, and convalescence. Although an SS doctor was in overall command, he rarely appeared and it was staffed mainly by prisoners.[16] By concentration camp standards the hospital was good, but starved of medical resources.

Fritz was taken to a room in the general medical ward. He was half paralyzed, his arms useless and senseless, his backside welted and bleeding, and his whole body shot through with pain. A Czech doctor gave him some strong painkillers and massaged his arms.

After a while, Gustl Herzog came in with Erich Eisler and Stefan Heymann. All three regarded Fritz with both pity and foreboding. When the doctor had gone, they questioned him anxiously about what the Gestapo had wanted with him. Fritz described Grabner's accusations and the alleged escape plan.

"Did you tell him anything?" Stefan asked.

"Of course not. I don't know anything."

That answer didn't satisfy them any more than it had the Gestapo.

"Did you reveal any names—any at all?"

Fritz shook his head painfully.

Despite the state he was in, his friends interrogated him over and over: *Had he named any names at all?* No, he insisted; he'd told Grabner nothing. To their minds, it was suspicious that he'd been allowed back to camp. It was possible that Grabner hoped Fritz might unwittingly betray his accomplices somehow; or it could be simply that the cells in Auschwitz I's Death Block were overflowing (as they often were).

Eventually they were satisfied that Fritz hadn't betrayed them. They were safe—for now. But Stefan and Erich were positive that Grabner wouldn't let the matter end there. He would resume the interrogation the next day, and Fritz's torture would continue until he either confessed or died. Something had to be done.

For the time being, they had Fritz moved to the infectious diseases block, where typhus and dysentery patients were kept, adjoining the morgue in the farthest corner of the camp. The SS doctor and his medical orderlies rarely went in there. Fritz was put in an isolation room. So long as he didn't pick up an infection, he'd be safe for the time being. But he couldn't hide in there forever; in order to prevent a manhunt when he failed to show up at roll call the next morning, his name would have to be entered in the hospital records. And then the Gestapo would come for him. Whichever way they looked at the problem, they kept coming back to the only solution: Fritz Kleinmann had to die.

Accordingly, Sepp Luger, the camp senior responsible for hospital administration, recorded the death of prisoner 68629 in the register. No details were required; the register provided only a single line for each patient, with admission number, prisoner number, name, dates of admission and departure, and reason for departure. In this column there were only three options: *Entlassen* (discharged); *nach Birkenau* for those selected for the gas chambers; or a stamped black cross for the dead. Gustl Herzog ensured that Fritz's death was also recorded in the general prisoner records office.[17]

The truth would remain an absolute secret among the conspirators. The news that Fritz had died from his injuries had to be broken to his many friends. Not even Gustav could be let into the secret—the risk was too great—and so he was given the devastating, heartbreaking news that his beloved Fritzl had been murdered by the Gestapo. The grief was so great that Gustav couldn't bring himself to record it in his diary, which had lain untouched for weeks.

While Gustav mourned, the conspirators faced the pressing matter of what to do with the living, breathing Fritz. While he began to recover from his injuries, he was kept in isolation in the hospital. Each time an inspection was carried out by the SS doctor or his male nurse, Fritz was helped out of his bed by his old friend Jule Meixner, who worked in the hospital laundry, and hidden in the storeroom among the bundles of linen.

Fritz had no idea what would become of him. Watching the dysentery patients drag themselves to the latrine buckets in the outer room and the typhus patients writhing feverishly in their sweat-soaked beds, he knew he couldn't stay in this place much longer, injuries or no injuries.

Word eventually came through from the Monowitz Gestapo that Grabner had dropped the investigation because of Fritz's death. It was time to move on.

Fritz was given a new identity, taken from a deceased typhus patient. He didn't recall the poor man's name—only that he was a Jew from Berlin, a relatively recent arrival whose prisoner number was up in the 112,000s. It was impossible to erase Fritz's tattoo or give him a new one with the dead man's number, so they just bandaged his forearm and hoped that nobody would demand to see it. Stefan Heymann spent a lot of time with him, advising him on how they would need to proceed and the precautions they would have to take when assigning him to a labor detail.

It was all one to Fritz. Since his ordeal a lassitude had entered his soul and he no longer much cared whether he was discovered or not. The long grinding of grief, starvation, and hopelessness had worn down his resistance at last, and he had started to drift into the helpless mind-state that led to becoming a *Muselmann*. He confessed to Stefan that he was considering ending it all as soon as possible—it was so simple to rush the sentry line while on an outside work detail, or to throw oneself on the electrified fence in the camp. One gunshot—a single fleeting instant—and the pain and wretchedness would all be over.[18]

Stefan had no patience with these thoughts. "Can you imagine what killing yourself would do to your father?" he said. "He believes right now that his son is dead, but in time—perhaps soon—he'll learn the truth. Imagine if he were to discover that you've been alive all this time only when you commit suicide—just think of that."

Fritz had no argument against this. After all they had been through together, for Fritz not only to cave in to the SS but to allow them to finish him off—it was too much. "They cannot grind us down like this," his papa always said; endurance was everything, misery was only for a time, hope and spirit were undying.

Stefan promised to do everything he could to keep Fritz safe in the hospital. When he was well enough to work, a place would be found in some outside detail where he could remain unnoticed. The death rate and turnover of prisoners was such that few ever got to know the others well.

Fritz understood this and trusted Stefan with his life, but he had doubts. People knew his face—including some of the SS. And sooner or later his father must find out. At least seven men in the resistance knew Fritz's secret, and his papa was their friend as well. Gustav was prominent in the camp now, and his high profile would make the possession of so explosive a secret extremely dangerous for him.

After three weeks, Fritz was well enough to leave the hospital. His friends smuggled him to block 48, where the senior was Chaim Goslawski, a resistance member. His block was mainly populated by Germans and Poles who didn't know Fritz.

Next day, he went to work. A position as a warehouseman had been found for him in a different section of the locksmith detail. One of the kapos, a man named Paul Schmidt, was in on the secret, and kept an eye on him. Marching out through the gates each morning and back in the evening, Fritz went through suffocating terror, expecting to be recognized by a guard or a hostile kapo. He kept in the middle of the group, marching with his eyes fixed forward and his face expressionless while his heart pounded.

As the weeks passed and nobody seemed to notice him, he began to feel more settled at work. For the time being, his secret appeared safe.

אבא

One evening Gustav was sitting in the dayroom of block 7 when one of his block-mates tapped him on the shoulder. "Gustl Herzog's outside," he said. "Wants to see you."

Gustav stepped outside and found his old friend in a state of suppressed excitement. *Follow me*, he indicated, and led Gustav down the path beside the building, away from the road. Behind the first row of barrack blocks stood some smaller buildings— latrines, the Gestapo bunker, and a small bathhouse. Herzog led Gustav toward the bathing block. A figure emerged from the gloom inside the doorway, whom Gustav recognized as the bath supervisor, a young Buchenwald veteran who had been a friend of Fritz's. He glanced around and, seeing that the coast was clear, gestured for Gustav to go inside.

Wondering, Gustav entered the building alone, inhaling the familiar smell of musty, soapless damp. In the low light, he saw the outline of a man standing back in the shadows of the boiler room. The figure came forward, his features resolving into the face of Fritz.

It was unbelievable, miraculous. For Gustav, who made it an article of faith never to abandon hope, no matter how desperate the circumstances, the astonishment was indescribable. To hold his son in his arms again, to inhale the smell of him, to hear his voice, was beyond hope, beyond everything.[19] Their survival had not been in vain after all.

After that first reunion, they met whenever they could, always at night in the bathhouse. Now that his grief and loss were taken away, Gustav's mind was invaded anew by all the cares of fatherhood, redoubled now that Fritz was in so much more danger than he'd ever been before. Gustl Herzog and the others assured him that they were doing all they could to safeguard him, but would it be enough?

אב ובן

In the autumn, some marvelous news came from Auschwitz I. The SS had suddenly removed Maximilian Grabner from his post as head of the camp Gestapo.

It was more than just a dismissal. For a long time, there had been questions in Berlin about Grabner's conduct. Even by SS standards the number of deaths he ordered raised an eyebrow—not so much at the scale of the murder but the disorderly way in which it was done. In Himmler's mind, the Final Solution—and killing generally—was an industrial business, to be conducted cleanly, efficiently, and systematically. It wasn't a game or a personal fetish. Grabner's sadism and bloodthirstiness placed a black mark against his name. However, his fall was ultimately brought about by his corruption.

Like many senior concentration camp officers Grabner had used his position to enrich himself with the valuables taken from Jews murdered in Birkenau, which were intended for SS profit. Unlike most, he had done so on a colossal scale, sending home whole suitcases crammed with misappropriated loot. The scale of his corruption had prompted an SS investigation. He was suspended from his post and placed under arrest along with several accomplices, including the insouciant mass-murderer Gerhard Palitzsch.[20] Rudolf Höss, commandant of Auschwitz, who had aided and abetted Grabner, was also removed.

The new commandant, Arthur Liebehenschel, took over in November 1943.[21] He initiated a shake-up of the whole Auschwitz complex; staff were replaced and order and discipline were imposed more firmly on the SS.

What was important to Fritz was that the gravest threat to his safety had been unexpectedly lifted. Grabner was gone, and amid all the turmoil there was little chance of one prisoner in Monowitz being taken much notice of by the Gestapo. Shortly afterward, on the night of December 7, a fire broke out in the Gestapo building at Auschwitz I, destroying the records of Grabner's misdeeds.[22]

Eventually, as obscurity settled over the whole Grabner episode and the need for concealment faded, Fritz Kleinmann came quietly back to life. His entry in the camp register was reinstated, and the Berlin Jew who had died of typhus was forgotten.

But although the need for absolute secrecy had passed, Fritz still had to be careful; if he were noticed by any SS guards who had been aware of his death—especially the Gestapo sergeants Taute and Hofer—there would be trouble. But among the thousands in Monowitz, and the hundreds of thousands who entered and were transferred back and forth between the Auschwitz camps, and the tens of thousands murdered, who would take notice of one prisoner's discreet resurrection?

As winter came on, Gustav used his position to have Fritz transferred to join him in the VIP block. Now they could be together in the evenings without resorting to risky meetings outdoors. It was a socially tricky situation; because of his low status, Fritz wasn't permitted to sit in the block dayroom when his papa went there to talk with his friends; instead he had to sit on his bunk alone, which was also technically illegal as bunks were for sleeping only.

Still, it was warm and safe. It was certainly better than the place he'd been in before his death; his block senior, a man named Paul Schäfer, hadn't been able to stomach the stench of men's bodies in the bunk room and had kept all the windows open in all weathers. Simply for sadism's sake, he also turned off the heating, so the men's damp clothes wouldn't dry. If anyone was caught trying to keep warm by sleeping in his uniform, Schäfer would beat him up and confiscate his rations.

"And so the year of 1943 goes by," Gustav wrote. Winter was upon them again; snow began to fall and the ground hardened. This would be his and Fritz's fifth winter since being taken from their home, their fifth year of relentless nightmare. And yet, as much as they had endured and suffered so far, the worst was yet to come.

15. The Kindness of Strangers

אחים

"Catch!"

Fritz leapt in the air, stretching for the ball as it sailed over his head; it bounced off one of the empty market stalls and skittered into the road. He ran and whipped it up, glancing up to see a policeman coming round the corner into Leopoldsgasse. The constable stared hard at him, and Fritz stood up straight, hiding the ball—really a tightly wrapped bundle of rags—behind his back. Soccer wasn't allowed on the streets. When he'd gone, Fritz turned and ran back into the market, dropping the ball and kicking it toward his friends.

It was the end of the day and the last of the farmers were clearing away their unsold wares. They mounted their carts and, chucking the reins, clopped off along the street. Fritz and his friends ran among the empty stalls, tossing the ball back and forth. Only Frau Capek the fruit seller was still at her post; she never packed up until it got dark. In the summer she would give the kids corn cobs. A lot of them were poor and would take all the free leftovers they could get—ends of sausage from the butcher, bread crusts from Herr König at the Anker bakery, whipped cream from Herr Reichert's cake shop in the Grosse Sperlgasse, just round the corner from the school.

Fritz caught the ball as it came his way, and was about to toss it back when they all heard the distant, familiar hooting of horns: *ta-raa ta-raa.* The fire engine going out on a call! In a welter of excitement they ran, dodging among the passersby—the late housewives with their shopping, the Orthodox Jews in their black coats and beards hurrying home for the start of Shabbat before the light began to fade. "Wait!" Fritz turned, and saw the little figure, legs pumping, running after him. Kurt! He'd forgotten all about him. He waited for his younger brother, and by the time he caught up, his friends were out of sight.

Kurt was only seven—a generation apart from Fritz, who was fourteen, but they were close. Fritz regularly let him tag along, learning their games and the ways of the streets. Kurt had his own gang of little pals, and Fritz's gang acted as their guardians.

They passed old Herr Löwy, who'd been blinded in the Great War, trying to cross the street, which was busy with trucks and heavy wagons from the coal sellers and breweries, clattering along pulled by massive Pinzgauer horses. Fritz took the old man's hand, waited for a gap, and helped him across. Then, beckoning Kurt to follow, he took off after his friends.

They caught up with them coming back along Taborstrasse, their faces streaked with cream and icing sugar. They hadn't found the fire, but they'd passed by Gross's confectioner's and bagged a ton of leftover cakes. Fritz's schoolfriend Leo Meth had saved a cream slice for Fritz, who divided it with Kurt.

Cheeks bulging with pastry, they walked back toward the Karmelitermarkt, Fritz holding Kurt by his sticky, sugary hand. Fritz enjoyed the comfort of comradeship; the fact that some of his friends were different, that while his parents neglected to go to synagogue, their parents stayed away from church, or that Christmas meant something more to them than it did to him— these things seemed of no significance, and the thought that he and Leo and the other Jewish kids might ever be divided from their friends by these trivial things never crossed their minds.

It was a warm evening, and tomorrow was Saturday—perhaps they'd go swimming in the Danube Canal. Or they might join with the girls to play theater in the basement of number 17. Frau Dworschak, the building supervisor, whose son Hans was one of Fritz's playmates, often let them illuminate the place with candles, and Herta and the other girls would put on a fashion show, dressing up in scavenged clothes and parading up and down like models. Or they'd all do a version of *William Tell* for an audience who paid two pfennigs each for admission. Fritz loved those burlesques.

He and Kurt reached home in the warm summer dusk. Today had been a good day in an unbroken string of good days. The kids of Vienna picked their joy from the streets like apples from a tree; all one had to do was reach up and it was there for the taking. Life was outside of time, inviolable.

בן

Fritz was torn from a pleasant dream by the shrill screech of the camp senior's whistle. His eyes opened into darkness, and his nostrils woke to the stench of three hundred unwashed bodies and three hundred musty, sweat-soured uniforms. His brain, startled out of bliss, registered the shock of his situation, as it did every predawn morning.

The man in the middle bunk climbed down and pulled on his jacket, along with the dozen others who were on coffee duty. Fritz wrapped his blanket tightly about him and closed his eyes, settling into the straw mattress and chasing the tatters of his dream.

An hour and a quarter later he was woken again by the bunkroom lights flicking on. "All up!" barked the room orderly. "Up, up, up!" In an instant the three-tiered bunks sprouted legs, arms, bleary faces, clambering, treading on one another, pulling on striped uniforms. Fritz and his papa took down their mattresses, shook them out, then folded their blankets and laid it all straight. After the men had splashed and scrubbed their faces in cold water in the washhouse—jam-packed with the inhabitants of the six surrounding blocks—and polished their shoes from the barrel of greasy boot polish scavenged from the Buna Werke, they lined up in the bunk room for their acorn coffee, brought in in huge eight-gallon thermos canisters. They drank it standing up (sitting on the bunks was forbidden). Those who'd managed to save a bit of bread from the evening before ate it now, washing it down with the sweet, lukewarm coffee. The orderly inspected their bunks, uniforms, and shoes.

The atmosphere was more convivial than in any block Fritz had been in before. The *Prominenten* of block 7 looked after themselves well.

At five forty-five, still in darkness, they trooped outside and formed rows in front of the building. All along the street, prisoners were spilling out of their barracks to be counted by their block seniors. Not even the sick or the dead were excused— usually each block would produce at least one or two corpses each morning. They were carried out and laid down to be counted with the rest.

The thousands of prisoners marched along the street and wheeled into the roll-call square, lit by floodlights. They formed orderly ranks, each man with his assigned place within his block, each block in its assigned place among the others. The sick and the dead were carried along and put at the back.

SS Blockführers prowled up and down the columns, looking for men out of place, lines not straight, counting up the men from their block, taking a tally of the dead. Any infraction of perfect drill—especially if it led to a counting error—resulted in a beating. When the Blockführers were satisfied, they took their reports to the Rapportführer on his podium at the front. While the prisoners continued to stand motionless—however cold or wet the weather—he went meticulously through the whole count.

By the time Lieutenant Schöttl arrived on the square, they had been standing at attention for about an hour. Fritz watched warily as Schöttl took the podium; he was still afraid of being recognized and singled out, a fear that would never entirely subside.

Recent events had put him more on edge than ever. In September, during the final weeks of the Grabner regime, an informer had been discovered among the prisoners.[1] The Gestapo were constantly looking for subversive activities, and the resisters had to be constantly vigilant. A prisoner clerk in the Monowitz Gestapo had identified kapo Bolesław "Bolek" Smoliński—a bigot and anti-Semite with a particular loathing for communists—as an informer working for SS-Sergeant Taute.

This vital intelligence was discussed among the resisters. Curt Posener (known as Cupo), one of the old Buchenwalders, pointed out that Smoliński was friendly with the camp senior responsible for the hospital, a main nexus for the resistance. This was a terrible vulnerability. Cupo talked it over with Erich Eisler and Stefan Heymann. Eisler suggested that they try talking to Smoliński, to make him see the error of his ways. Stefan and Cupo argued strenuously against this dangerous idea. Nonetheless, Eisler disregarded the warnings and talked to Smoliński. The reaction was instantaneous—Smoliński went straight to the Gestapo. Immediately, Erich Eisler and Curt Posener were seized and taken to Auschwitz I, along with six others, including Walter Petzold and Walter Windmüller, both highly respected functionary prisoners and members of the resistance. They were put in the Death Block bunker and subjected to days of interrogation and torture. Smoliński was held with them.

Eventually Curt Posener and one other returned to Monowitz, battered and physically broken. Like Fritz, they had resisted the torture and given away no information. Smoliński was also released and resumed his position. Walter Windmüller succumbed to his injuries and died in the bunker. Poor Erich Eisler, having outed himself as a resister by talking to Smoliński in the first place, was taken to the Black Wall and shot.[2] Eisler had dedicated himself utterly to people's welfare; even before becoming a prisoner himself, he'd worked for the Rote Hilfe (Red Aid), a socialist organization that provided welfare to prisoners' families.[3] In the end, it had been his humane temperament that was his downfall, thinking he could talk a man like Smoliński into behaving decently.

"Attention! Caps off!" yelled a sergeant's voice over the loudspeakers, and five thousand hands whipped off their caps and folded them neatly under their arms. They stood at attention while Schöttl checked through the assembled lists of prisoners, noting new arrivals, deaths, selections, and assignments.

Finally: "Caps on! Work details, move!"

The parade dissolved into chaos as each man moved to his allotted detail, coalescing into units counted off by their kapos. They marched along the street to the main gate, which swung open. Many were weak and lethargic, having reached the last of their strength; before long they would be selected for Birkenau or be among the corpses brought out to be counted at roll call.

As the columns passed, the prisoner orchestra, in their little building beside the gate, played stirring tunes. The musicians were led by a Dutch political, with a German Roma on violin; the rest were Jews from various nations. It struck Fritz that they never seemed to play German tunes—only Austrian marches from the days of the empire. His papa had once marched to these tunes on the parade grounds of Vienna and Cracow, and gone to war accompanied by the same martial airs. The camp orchestra had good musicians, and sometimes on a Sunday Schöttl permitted them to put on a concert for the more privileged prisoners. It was a surreal sight—the motley musicians playing classical music to an audience of standing prisoners, with SS officers in chairs to one side.

The sky was growing light as they marched along the road toward the checkpoint at the gates of the Buna Werke, each column guarded by an SS sergeant and sentries. Depending on where in the factory complex they worked, some of them had up to two and a half miles still to march, and then a twelve-hour shift and a two or three mile march back to another several hours of roll call in the floodlit cold and rain.

Fritz went to his work in the warehouse and began another dull but safe day of moving stock about. He had no way of knowing that today would mark the beginning of a transformation in his existence.

He was chatting to another Jewish prisoner when one of the German civilian welders who happened to be nearby broke in on their conversation.

"Nice to hear German spoken," he said. "I haven't come across many Germans since I came to work here—most of you lot are Poles or other foreigners."

Fritz looked at the man in surprise. He was fairly young, and moved with a lame, halting gait.

The man glanced at their uniforms. "What are you in for?" he asked.

"Excuse me?"

"What crime?"

"Crime?" said Fritz. "We're Jews."

He had to repeat himself several times to get the concept across. The man was mystified. "But the Führer would never lock up anyone who hasn't done anything wrong," he said.

"This is Auschwitz concentration camp," said Fritz. "Do you know what Auschwitz stands for?"

The man shrugged. "I've been in the army, on the Eastern Front. I've got no idea what's been going on at home."

That explained his lameness; he'd been wounded and invalided out.

Fritz pointed to his badge. "This is the *Judenstern*—the Jewish star."

"I know what it is. But you don't get put in a camp just for that."

This was incredible, and vexing. "Of course we do."

The man shook his head in disbelief. Fritz's temper began to fray. Such blindness was stupefying—the man might have missed the escalation since 1941 while he was at the front, but where had he been since 1933 when the persecutions began, or 1938 when Kristallnacht happened? Did he believe that Jews had all emigrated of their own free will?

It wasn't safe to argue with a German, so Fritz gave up trying to convince him.

Later that day, the man approached him again. "We've all got to pull together, you know," he said. "We all have to defend the Fatherland and work for the common good—even you lot have got your part to play."

Fritz bit his tongue. The man rambled on about attitude and duty and the Fatherland until Fritz couldn't stand it any longer. "Can't you see what's happening here?" he said angrily, gesturing to take in the factories, Auschwitz, the whole system. Then he walked away.

The civilian, perplexed by Fritz's attitude, wouldn't leave it alone. All that day he kept coming up to Fritz. Duty and Fatherland were his repeated themes, and how prisoners must be prisoners for a good reason. But, despite his persistence, with each repetition he sounded less sure of himself.

Eventually he fell silent, and for the next few days went about his welding without speaking. Then one morning he approached Fritz, surreptitiously passed him a chunk of bread and a large stick of sausage, then walked off.

The bread was half a loaf of *Wecken*, an Austrian bread made from very fine flour. Fritz tore off a piece and put it in his mouth. It was blissful; nothing like the military *Kommisbrot* they were given in the camp. This was a taste of home and heaven, bringing back memories of the morsels he and his friends used to get at the close of day from the Anker bakery. He squirreled it away, along with the sausage, intending to smuggle them back to camp to share with his father and friends.

An hour or two later, the civilian passed his way again and stopped. "There aren't many Germans here," he said. "It's nice to have someone to talk to." He hesitated, and there was a troubled look on his face that Fritz hadn't seen before. "I saw something," he said awkwardly. "This morning on my way in to work . . ." Visibly upset, he haltingly described how he'd seen the corpse of a prisoner hanging on the electric fence of the Monowitz camp. Even as a veteran of the eastern front and no stranger to atrocity, he'd been shaken. "They tell me it was suicide. They say it happens now and again."

Fritz nodded. "It happens pretty often. The SS leave the bodies up for a few days to intimidate the rest of us."

The man's voice shook with emotion: "This is not what I fought for," he said. There were tears in his eyes. "Not that. I want nothing to do with that."

Fritz was astounded—a German soldier in tears over a dead concentration camp inmate. In Fritz's experience, Germans—whether they were soldiers, police, SS, green-triangle prisoners—were all of a kind. The only exceptions were the socialist political prisoners: otherwise they were callous, bigoted, and brutal.

The man began to tell Fritz his story. His name was Alfred Wocher. He was Bavarian-born, but married to a Viennese woman, and his home was in Vienna—hence the *Wecken* loaf. Fritz didn't mention that he was from Vienna; instead he listened while Wocher talked about serving in the Wehrmacht on the eastern front, how he'd been awarded the Iron Cross and reached the rank of sergeant. After being severely wounded he'd been sent home on indefinite leave; he hadn't been discharged from the army, but he would never again be fit for active service. As a skilled welder he'd been sent to IG Farben to do civilian work.

It occurred to Fritz that Wocher might be a useful contact. Back in camp that evening, he went to the hospital to talk it over with Stefan Heymann; he described Alfred Wocher and repeated everything he'd said. Stefan was unsettled by the whole thing. He advised Fritz to be careful—you couldn't trust Germans, especially not a veteran from Hitler's army. After Smoliński, the resistance was more wary than ever about potential informers. And the last time Fritz had become friendly with a civilian it had nearly cost him his life—not to mention putting his friends and his father through a world of grief.

Fritz understood all too well. He knew that Wocher should not be trusted. But somehow—maybe it was the Viennese bread or his obviously genuine upset over the dead prisoner—somehow Fritz couldn't help it. He went back to work and, in defiance of Stefan's advice and his own good sense, continued talking with the old soldier.

It was hard to avoid him. Wocher came to Fritz, usually because he wanted to get something off his chest, some question or other about Auschwitz. To Fritz it seemed suspiciously like probing, and the sensible thing to do would be to turn his back and refuse to even listen. But he answered, without going into detail, with facts about Auschwitz. Wocher would bring copies of the *Völkischer Beobachter*, the Nazi Party newspaper, to show Fritz what was going on in Germany. Fritz didn't mind—newspaper had a value in the camp, and it had to be said that wiping the asses of Jews was as good a use for the *Beobachter* as one could imagine. More welcome were the gifts of bread and sausage. One day, out of the blue, Wocher offered to convey letters for Fritz. If there was anyone in the outside world he wanted to communicate with, Wocher would get messages to them.

So there it was—entrapment. Or so it seemed. The temptation to get in touch with relatives still in Vienna—and hopefully find out what had become of his mother and Herta—was extremely strong. Fritz's instinct was to test the man in some way. But to what purpose? If Wocher was a Nazi informer, what good would it do to prove it? He'd still end up in the bunker.

Fritz discussed the matter again with Stefan Heymann. Knowing that Fritz would always go his own way, Stefan told him that it was up to him alone; he couldn't help him with this.

Not long afterward, Wocher happened to mention that he was about to go on leave. Here was Fritz's opportunity; Wocher had mentioned that he'd be passing through Brno and Prague en route to Vienna, so next day Fritz came to work with a couple of letters directed to fictional addresses in both Czech cities, claiming he had family there. Wocher took them happily, promising to deliver them personally. (He wouldn't entrust them to the postal system, which was subject to spying.) Fritz guessed that if Wocher was false, he naturally wouldn't bother trying to deliver the letters and wouldn't discover that the addresses weren't real.

When Wocher reappeared at work a few days later, he was livid. He'd tried to deliver both letters, and been unable to find either address. He assumed Fritz had duped him just to make a fool of him, and was hurt as well as angry. Fritz apologized, concealing his delight and relief; he was now almost sure that Wocher wasn't a provocateur.

He began to reveal more about what Auschwitz really was, describing how Jews came in transports from Germany,

Poland, France, the Netherlands, and countries in the east; about the selections in Birkenau: how children, the old, the unfit, and most of the women were sent to the gas chambers, while the others became slaves. Wocher had seen glimpses of it for himself; now he understood the long trains of closed wagons he'd seen coming in along the southeastern railway past Monowitz, heading toward Oświęcim. On the factory floors he'd heard civilians talking about such things, and was beginning to realize that he'd missed a lot while at the front.[4]

It was hard to miss what was going on. Like a metastasizing cancer, Auschwitz was spreading and growing. Sweeping changes and expansions had been initiated, and Auschwitz III-Monowitz was now the administrative hub of a growing number of subcamps pustulating throughout the countryside around the Buna Werke. A commandant had been installed above camp director Schöttl, a pallid, blank-eyed captain called Heinrich Schwarz, who liked to participate in beating and murdering prisoners, working himself into a foaming rage in the process. Schwarz was devoted to the Final Solution, raging against Berlin whenever there was a lull in the flow of Jews into Auschwitz.[5]

New transports for the IG Farben subcamps sometimes came directly to Monowitz now, and for the first time Fritz witnessed with his own eyes what he had previously only heard about—the bewildered people herded from the freight wagons onto the ground near the camp, loaded down with luggage. Men, women, and children, thinking they had come to be re-settled.[6] Many were frightened, others happy and relieved to find friends again among the mass after days in the suffocating wagons. The healthy men were separated and marched to the camp, while the women, children, and elderly were put back on the train, which rolled on to Birkenau.

In Monowitz the men were made to strip in the roll-call

square. Many tried to keep hold of precious possessions, but they were nearly always found. Everything was taken to the storage block known as "Canada" (which was believed to be a land of riches) for sorting and searching. The prisoner detail responsible for handling the plunder, working under close SS oversight, searched through it like prospectors panning dirt, teasing open seams to look for concealed valuables.[7]

Fritz took a particular interest in newcomers arriving from the ghetto at Theresienstadt, many of whom were originally from Vienna. He begged them for news of home, but they had little to tell. More up-to-date news came when deportations directly from Vienna began arriving. Virtually all the registered Jews had gone from the city now, and the Nazi authorities had begun deporting *Mischlinge*—people who were born from the intermarriage of Jews and Aryans and were therefore both and neither. Frustratingly, nobody could tell him anything about his remaining relatives and friends, if any were still alive.

When Alfred Wocher mentioned that he was shortly going to Vienna on leave, Fritz saw his chance. He felt he could trust him now, and hoped the trust was reciprocated. Fritz gave him the address of his Aunt Helene, who lived in Vienna-Döbling, an affluent suburb across the Danube Canal from Leopoldstadt. She had married an Aryan and been baptized a Christian. Helene's husband was now an officer in the Wehrmacht, and so far she had remained secure from the Nazis. Her son, Viktor, was the cousin from whom Kurt had acquired his hunting knife. Rather than write a letter, Fritz gave Wocher a verbal message—simply informing her that he and his papa were still alive and well, and asking her to pass the news on to any other surviving relatives. Wocher took the address and set off.

He returned a few days later. The mission hadn't been much more fruitful than the previous one. The address had been genuine enough, but the lady who'd answered the door to him had been decidedly unfriendly—she'd denied all knowledge of anyone by the name of Kleinmann, and slammed the door in Wocher's face.

Fritz was mystified, and questioned Wocher closely. Was he sure he'd gone to the right address? Eventually he discovered what had gone wrong. What he hadn't realized was that when he was away from the factory, Alfred Wocher reverted to army uniform. His appearance on Aunt Helene's doorstep, asking about her Jewish relations, must have scared the poor woman out of her wits. In truth, it was worse even than Fritz guessed. Helene's officer husband had died in the war, and she felt terrifyingly exposed without the protection his status had given her.

One thing at least had come of it: Fritz now trusted Alfred Wocher completely.

When Christmas came and Wocher set off to Vienna again for the holiday season, Fritz provided him with some more addresses of some non-Jewish friends of his papa's from the neighborhood around the Karmelitermarkt. He also gave him the address of the old apartment in Im Werd, and a letter for his mother.[8] Despite everything, Fritz couldn't give up hope completely. He needed to believe that she and Herta were alive and well. Somebody must know.

חברים

Leopoldstadt had lost its heart. Formerly Jewish shops were still untenanted, businesses boarded up, homes empty. When Alfred Wocher ascended the stairs of the apartment building at Im Werd 11, half of the apartments were unoccupied.[9] So much for the Nazi claim that Jews were taking up scarce living space that was needed for true Germans.

There was no answer when he knocked at apartment

number 16. The door had probably never been opened since Tini Kleinmann turned the key in the lock in June 1942. Enquiring at some of the other apartments, Wocher eventually came across a man named Karl Novacek, who had been a friend of Gustav's. Karl, who worked as a cinema projectionist, was one of the handful of non-Jewish friends who had remained loyal to the Kleinmanns throughout the Nazi persecutions.[10] He was overjoyed to learn that Gustav and Fritz were still alive.

He wasn't the only one. There were other true friends in the same street—Olga Steyskal, a shopkeeper who had an apartment in the building next door, and Franz Kral, a locksmith. Their reaction was the same as Karl's. As soon as they heard the news, the three friends hurried across the street to the market and returned with baskets of food for Wocher to take back to Auschwitz. Word also reached Fritz's cousin Karoline Semlak—Lintschi as she was better known—who lived a few streets away. Lintschi had become a Christian Aryan by marriage, but unlike poor Aunt Helene in Döbling, she had no qualms about exposing her Jewish connections. She put together a package of food and wrote a letter, enclosing photographs of her children. Olga—or Olly as her friends called her—also wrote Gustav a letter; she'd always been deeply fond of him, as he was of her; there might have been sparks between them if he hadn't already been married.

It was an incongruous, improbable occasion: a group of Aryan friends and a converted Jew packing off a Bavarian soldier in Wehrmacht uniform with armfuls of loving gifts for two Jews in Auschwitz. It was strangely beautiful, but it left Wocher with a problem: the food parcels filled two suitcases. Conveying it all safely to Fritz was going to be quite a challenge.

Back at Auschwitz, he smuggled the gifts into the factory in stages and passed them over. The food was very welcome, but even more precious to Fritz was the news of Lintschi and their friends. He asked eagerly about his mother and sister, but Wocher shook his head. Everyone he'd spoken to had said the same—Tini Kleinmann and her daughter had gone with the deportations to the Ostland and never been heard from since. Fritz's disappointment was bitter. But he still clung to the faint possibility that they weren't dead. His aunts, Jenni and Bertha, had been deported on one of the last transports to leave Vienna for Minsk the previous September. Jenni had no family of her own other than her talking cat, but Bertha had left behind her daughter, Hilda (who was married to a non-Jew), and grandson.[11]

Sharing most of the food among his workmates, Fritz took the rest, along with the letters, back to his papa. Despite the crushing news about Tini and Herta, Gustav was heartened to hear from his dear friends. His nature rebelled against giving up hope, and it gave him joy to think that he would be able to write to people he loved.

There was a much bleaker reaction from Gustl Herzog and Stefan Heymann when Fritz told them what he'd done; despite his own confidence in Alfred Wocher's trustworthiness, Stefan in particular was deeply suspicious. He warned Fritz against any further involvement with the German.

Fritz kept his own counsel. His respect for Stefan was great, but his longing for the old world and his family was greater.

16. Far from Home

אבא

"Dearest Olly," Gustav wrote.

> Your kind letter to me is received with many thanks, and you must forgive me for leaving you for so long with no word from me and Fritzl, but I have to take great care not to cause any trouble for you. For your kind package I thank you many times over . . . It makes me so glad that I have such kind and good friends when I am so far from home.[1]

It was the third day of the new year of 1944, and there was a faint whiff of hope in the air. Gustav's pencil darted rapidly across the square-ruled sketch paper.

> Believe me, dear Olly, through all the years I have always recalled the beautiful hours that I spent with you and all your dear ones, and have never forgotten you. As for me and Fritzl, the years have been hard, but I owe it to my will power and energy that it was always my choice to keep going.
>
> If it should be granted to me to be in contact again with you and your dear ones, it will make up for what I have been missing—that for two and a half years I have had no news about my family . . . But I'm not letting my hair turn gray over it, because someday I will be reunited with them. As far as I am concerned, dear Olly, I am still the old Gustl, and intend to stay that way . . . Anyhow, be assured, my dear, that wherever I am

I am always thinking of you and all my dear friends—now I close with the fondest wishes and kisses. Your Gustl and Fritz.

Gustav folded the sheets and put them in an envelope. Fritz would smuggle it into the factory the next morning and pass it to this German friend of his. Once again his boy had surpassed himself for courage and initiative. There was no restraining him; all Gustav could do was hope he didn't bring trouble on himself again.

As the weeks went by, Fritz took letters to Fredl Wocher from other Viennese prisoners—mostly Jews with Aryan wives at home. They took care to write in a way that wouldn't incriminate either the sender or receiver if they were intercepted by the Gestapo.

בן

Passing letters wasn't the only way in which Fritz subverted the system to benefit his comrades. Bonus trading was another.

Auschwitz had recently introduced bonus coupons for exemplary workers. Available only to non-Jewish prisoners in high-status occupations,[2] they could be exchanged for luxury items like tobacco or toilet paper at the prisoner canteen. The system—which was Himmler's idea—was meant to increase productivity, but in practice kapos often used the coupons as a means of rewarding special favors rather than good work.[3]

For many, the main attraction of the coupons was that they could be used to pay for visits to the camp brothel. This facility was another of Himmler's initiatives to reward productivity. It stood surrounded by a barbed-wire fence near the camp kitchens, and was known euphemistically as the "women's block."[4] The women were prisoners from Birkenau—German, Polish, Czech; none of them Jews—who had "volunteered" on the promise that they would be given their freedom in due course. There was a waiting list for customers at the brothel, and only Aryan prisoners with bonus coupons could apply. On admission, the customer was given an injection against venereal disease and an SS man assigned him a woman and a room. During the day, when the brothel was closed, the women could sometimes be seen taking walks outside the camp, each escorted by a Blockführer.

As an official Aryan, Gustav received bonus coupons, but had little use for them. Camp director Schöttl, who had perverse tastes, got vicarious thrills from listening to prisoners' detailed descriptions of their activities with the women; but although he tried several times to persuade Gustav to go to the brothel, he always declined, ruefully pointing out his advanced age. (He was only fifty-two, but by camp reckoning that made him a veritable graybeard; almost nobody lived that long.)

Since he didn't smoke either, Gustav had no particular need for his coupons, so he passed them to Fritz, who traded them on the black market.

Fritz had cultivated the acquaintance of the kapos in charge of the kitchen and the Canada store where belongings looted from prisoners were kept; both men were deeply corrupt, and both were addicted to the brothel. In return for coupons, Fritz received bread and margarine from the larder and valuable clothes from the Canada store—pullovers, gloves, scarves, and anything warm. He took his bounty back to the block and shared it with his father and friends.

He was uncomfortably conscious that his trade depended on the exploitation of the women in the brothel. In such a hostile environment, one person's philanthropy had to come at the cost of another's suffering. Eventually the women were replaced by a new batch of younger Polish girls. The original group, who had endured months of degradation on the promise of freedom, were sent back to Birkenau. They were never set free.[5]

✡

In the spring and early summer of 1944, the character of Auschwitz began to alter noticeably. Gustav recorded in his diary that Monowitz was receiving a constant stream of new prisoners, nearly all young Hungarian Jews. They brought with them a hollow-eyed melancholy, as well as news from the east that, by Gustav's reckoning, indicated that the war was going very badly for the Germans.

In March, Germany had invaded its former ally Hungary. Alarmed by the steady crumbling of Germany's forces on the eastern front and the likelihood of an Anglo-American invasion of northwest Europe, the Hungarian government had begun making secret overtures of peace to the Allies. In German eyes, this was a devastating betrayal. Hitler responded with swift fury, invading the country and taking control of its army.

Hungary had a population of around 765,000 Jews.[6] Their lives had been blighted by exclusion and anti-Semitism, but thus far relatively few of them had been harmed.[7] Now, in an instant, they were cast into the pit.

Systematic persecution began on April 16—the first day of Passover, the traditional celebration of divine liberation from bondage.[8] Einsatzgruppe units, reinforced by the Hungarian gendarmerie, began herding hundreds of thousands of Jews into makeshift camps and ghettos. It was rapidly, efficiently, and savagely done; the SS sent its two most experienced officers to take charge: Adolf Eichmann, who had developed his expertise in deporting Jews from Vienna; and Rudolf Höss, the former commandant of Auschwitz.

The first transports left Hungary for Auschwitz at the end of

April, containing 3,800 Jewish men and women. On arrival, most went to the gas chambers.[9] They were the first trickle in a human flood. To heighten efficiency, the "old Jew-ramp" at Oświęcim was replaced by a rail spur hastily laid right into the Birkenau camp, with an unloading ramp a quarter of a mile long.

Gustav later became acquainted with some of the Hungarian women who arrived in Auschwitz at that ramp, and they described to him in detail what happened.

On Tuesday May 16, the entire Birkenau camp was put on lockdown. Prisoners were shut in their blocks under guard. The only exceptions were the Sonderkommando and, incongruously, the camp orchestra. Shortly afterward, a long train came steaming and squealing along the rail tracks, through the archway in the brick gatehouse, and rolled to a halt at the ramp. The doors slid open, and from each wagon about a hundred people spilled out. Old and young, women, men, children, infants. Scarcely any had the faintest idea what manner of place they had come to, and many disembarked with light hearts, tired and disorientated but hopeful.[10] As the striped uniforms of the Sonderkommando moved among them, they weren't afraid. The sound of music from the orchestra added to the atmosphere of harmlessness.

Then came the selection. Everyone over fifty years of age, anyone who was lame or sick, children and their mothers, pregnant women, all were sent to one side. Healthy men and women between sixteen and fifty years old—about a quarter of the total—were sent to the other. As the day wore on, two more trains came in from Hungary. Two more selections, thousands of souls sent to left or right. Those designated fit for labor were labeled "Transit Jews" and sent to a section of the vast camp. The others were herded onward to the low buildings among the trees at the end of the rail tracks where foul-smelling smoke streamed from the chimneys night and day.[11]

Around fifteen thousand Hungarian Jews entered Birkenau that day; the exact number murdered would never be known, because not one of them—the dead or the enslaved—was ever registered as a prisoner of Auschwitz or received a number.[12] Even those assigned to the labor camps were not intended to survive long.

It was the beginning of a monstrous escalation that would mark the zenith—or rather the nadir—of Auschwitz as a place of extermination. Between May and July 1944, Eichmann's organization sent 147 trains to Auschwitz.[13] They arrived in Birkenau at the rate of up to five a day, overwhelming the system. Additional gas chambers that had lain dormant were put back into use. Four in all operated round the clock. Nine hundred overworked, traumatized Sonderkommando operatives herded the panicked women, men, and children naked into the gas chambers and hauled out the corpses afterward. The Canada detail filled block after block with looted clothes, valuables, and suitcases of belongings. The crematoria couldn't cope with the sheer number of dead, and pits were dug in which to burn the bodies. The SS went into a frenzy; so great was the rush to murder each newly arrived batch that gas chambers were often opened up while some victims were still breathing; those who moved were shot or clubbed; others were flung into the fire pits still half alive.[14]

Many of the men and women who passed the selections were sent to Monowitz. Gustav watched them arrive with bleak sympathy. "Many of them no longer have parents, because the parents are left behind in Birkenau," he wrote. Only a minority were like himself and Fritz—a father and son together, or a mother and daughter. Would they have the strength and luck to survive as he and Fritz had? Looking at their broken state, it seemed unlikely. Many already showed the vacant depression symptomatic of the transformation into *Muselmänner*. "Such a sad chapter," Gustav wrote.

אבא

By the middle of 1944, Gustav's upholstery detail had moved to premises in the Buna Werke. Such was the level of influence he now enjoyed that he'd been able to have Fritz transferred to work under him.[15]

The early months of the year had been tough: a savage winter with thick snow and outbreaks of fever and dysentery. Both of them had fallen sick and spent time in the hospital, in constant danger of being selected for liquidation. Gustav had been the first to fall ill, admitted along with dozens of others in February. He was in for eight days, recovering just in time to avoid a selection in which several men who'd been admitted at the same time were sent to the gas chambers. Another outbreak of sickness in late March had put Fritz in the hospital for over two weeks.[16]

Now that he was based in the factory, Gustav was finally introduced to Fredl Wocher, their benefactor, who now had his and Fritz's complete trust.

For Fritz, being in his father's workshop meant a resumption of his apprenticeship in upholstery, interrupted by the Anschluss of 1938. They worked under a civilian master from Ludwigshafen. "He's all right," Gustav wrote, "and he supports us wherever he can. The man is anything but a Nazi."

The loyalties of Germans were coming under increasing strain as the war unfolded and they began facing up to probable defeat and the reality of what the Nazi regime had done. On June 6, the long-anticipated invasion of France by Allied forces began. Meanwhile, the Red Army pushed relentlessly from the east.

In July the Russians swept into the Ostland, encircling Minsk and capturing the region where the remains of Maly Trostinets lay. The small camp had been decommissioned and razed in October 1943, having served its purpose. On July 22, units advancing into eastern Poland captured the huge concentration camp of Majdanek on the outskirts of Lublin, the first large-scale camp to fall into Allied hands. They found it virtually intact, complete with gas chambers and crematoria and the corpses of its victims. Eyewitness descriptions flew around the world, appearing in newspapers ranging from *Pravda* to *The New York Times*. In the words of one Russian war correspondent, the horror of it was "too enormous and too gruesome to be fully conceived."[17]

Pressure was growing on the Allied governments—who already had quite detailed intelligence about the camps, including Auschwitz—to do something directly to help. There were calls for bombing raids against camp facilities and railway networks. Allied air commanders considered and dismissed the calls; it was not a viable use of their resources, they said, which were fully committed to strategic bombing and air support for the advancing armies. And that was that.[18]

However, the SS were acutely conscious that some of the camps were located next to strategic industrial facilities that were at high risk of bombing—such as the Buna Werke at Auschwitz, which was just within striking distance of Allied long-range bombers. The Auschwitz SS decided to implement some air-raid precautions.[19] Shelters were set up at the Buna Werke and a blackout policy was put into effect across the Auschwitz complex. The task of equipping the factories for blackout fell upon Gustav Kleinmann, who was taken off upholstery work and put in charge of manufacturing blackout curtains. He was given a workshop with sewing machines and a team of twenty-four prisoners, mostly young Jewish women—"all well-behaved and reliable folk." While Gustav's team made the curtains, Fritz assisted the civilian fitters who installed them.

Gustav worked under a civilian manager called Ganz, a socialist, who would stop by the workshop to chat and share his lunch. Ganz was quite different from some of his fellow managers in this part of the factory, who lived in awe and terror of the SS and insisted that the Führer knew what he was doing; a few were thoroughgoing Nazis who reported any fraternization to the senior engineer, another loyal Hitlerite.

Some of the Polish women from the neighboring insulation workshop would smuggle bread and potatoes to the Jewish prisoners in the blackout workshop. Where they got the food from was a mystery, because their own rations were hardly plentiful. Gifts also came from two Czech curtain fitters, who acted as messengers for Czech Jews, much as Alfred Wocher did for Fritz, taking letters to their friends in Brno and bringing back gifts of lard and bacon.

The generosity was great, but the quantities, in the face of the thousands who needed it, were tiny. All but the most Orthodox Jews received bacon and other non-kosher food with gratitude, having long let go of the stricter elements of their faith.[20] Some, like Fritz, had abandoned their religion altogether, finding it impossible to sustain a belief in a God who cared about the Jews.

The women in Gustav's workshop, having been inside Birkenau, told Gustav all about what was going on there. Four Hungarian tailors allotted to his curtain detail described the roundups in Budapest. It had been like a tornado, far quicker and more ferocious than in Vienna. Hungarian Jews, despite living under their own anti-Semitic government, had been allowed to keep the Shabbat and attend synagogue, and had convinced themselves that the persecution stories coming out of Germany were exaggerated. Then the Nazis came and they saw for themselves.

For nearly two years Gustav had absorbed the stories out of

Birkenau, but what was occurring now was a new level of barbarism. "The stench of the burning corpses reaches as far as the town," he wrote. Every day he saw the trains pass by Monowitz on the railway from the southeast, the wagons closed tight. "We know everything that's going on. They are all Hungarian Jews—and all this in the twentieth century."

בן

With Fritz helping, Schubert fixed the last curtain to the office window. He tried to explain to the manager how to use the curtains, but it was difficult; Schubert was an ethnic German from Poland and spoke German very badly.

He and Fritz packed their tools away. As they did so, one of the civilians passed some ends of bread to Schubert, with a nod at Fritz. Schubert took them discreetly, slipping them into Fritz's toolbox. Fritz heaved the stack of curtains onto his shoulder, and they moved on to the next building. They got on well together, despite the difficulties of communication. Schubert came from the town of Bielitz-Biala, where Gustav had worked as a baker's boy in the early years of the century. Fritz rather enjoyed being out and about—it was almost like a taste of freedom. Each day he and Schubert returned to the workshop with their toolboxes full of scraps of bread.

The next building on the list was close to the main factory gates, where there was a checkpoint manned by an SS corporal known to the prisoners as *Rotfuchs*—Red Fox—because of his flaming red hair and temper to match. As they were passing, Fritz noticed Rotfuchs staring in irritation at a group of Greek Jews standing idle inside the gates. Fritz could sense that something was about to happen, and slowed his pace. Rotfuchs's anger got the better of him; leaving his checkpoint, he marched up to the Greeks and started yelling at them to get back to work. None of them spoke German, and they had no idea what he was saying. He began battering them savagely with the butt of his rifle.

Fritz couldn't stop himself. He dropped everything and ran across, throwing himself between Rotfuchs and his victims. "You have to get back to the checkpoint," he said, pointing toward the wide-open gate. "Prisoners might escape."

Other SS men might have been brought up short by this reminder of their duty, even from a Jewish prisoner. But not Rotfuchs. His pale, blotchy face turned purple with fury. "I'll do as I please!" he screamed. There was an oily *schlick-clack* of his rifle being cocked, and the muzzle pointed at Fritz.

So this was it; after all these years, it would end here in a moment of blind rage, all for some prisoners he didn't even know.

At that instant, as Rotfuchs was about to pull the trigger, the rifle was pushed aside by Herr Erdmann, a senior engineer, who had been drawn by the noise. Without hesitation, Fritz turned on his heel and walked determinedly into a nearby materials store. He knew better than to hang around at the scene.

It could go either way; he might still be shot as a punishment, or at the very least be whipped. But it never came to that; Herr Erdmann lodged a formal complaint against Rotfuchs with IG Farben, and the corporal was transferred to a different posting. The prisoners of Monowitz never saw him again.

Erdmann's action typified many Germans' feelings. The little respect remaining for the Nazi regime was being eroded by the ever-worsening situation Hitler had brought upon Germany. Many Germans were afraid of what would become of them, and for those who worked in and around Auschwitz, the more they learned about what the SS had really done, the less they could stomach it.

With his ability to move around the Buna Werke on curtain-fitting duties, Fritz was able to meet up with Fredl Wocher often. On one occasion, he introduced Fritz to some friends in the Luftwaffe antiaircraft batteries stationed around the perimeter. They had more rations than they needed, and gave Fritz several cans of meat and fish preserves, jam and synthetic honey.

Gifts of food had become more important than ever. With Germany afflicted by shortages, all resources were being channeled to the military on the front lines; citizens at home were on short rations, and prisoners in the concentration camps got almost nothing. The number of *Muselmänner* increased, and deaths from sickness and starvation escalated, as did the selections for the gas chambers. There was a limit to how far the donated food could go, but at least it helped a few. Fritz and his better-fed comrades gave up all their camp-issued rations to those who were starving.

How to share the food among such a large number was a constant worry to Fritz, and he was haunted by the harsh choices it forced him to make. "If we were to share it among so many, for each it would be no more than a drop of water on a hot stone." Giving food to a *Muselmann*, so starved that one look told you he would be dead within days, seemed like a waste.[21]

Hardening his heart against the terminally weak and dying, Fritz gave his spare food to the young. There were three boys in his block, all of whom had lost their parents to the gas chambers. One was his old playmate from Vienna, Leo Meth, who had initially escaped the Nazis by being sent to France, only to fall into the net after the German annexation of the Vichy zone. Fritz gave them his share of ration bread and soup, as well as a portion of the sausage and other morsels donated by people at the factories. In his mind it was a return for the kindness he'd received from elders when he was a vulnerable sixteen-year-old in Buchenwald.

Gustav too did what he could for young and needy prisoners. One day when a batch of new arrivals were being entered on the register, he heard the name Georg Koplowitz called. Gustav's mother had once worked for a Jewish family with the name Koplowitz; she'd been fond of them and remained with them until her death in 1928. Intrigued, Gustav tracked down this young man, and discovered that he was in fact the son of the very same family, the sole survivor of the selection at Birkenau. Gustav took Georg under his wing, giving him surplus food each day and fixing him up with a safe position as a helper in the hospital.[22]

The circle of kindness was completed by British prisoners of war who were forced to work in the factories alongside the Auschwitz inmates. They came from camp E715, a labor subcamp of Stalag VIII-B. Despite being within the SS-controlled Auschwitz zone, they were prisoners of the Wehrmacht, and it was Wehrmacht guards who escorted them to work and guarded them. They received regular welfare packages via the International Red Cross, and shared some of the contents with the Auschwitz prisoners, along with news about the war picked up from the BBC on secret radio sets in their camp. Fritz particularly enjoyed their chocolate, English tea and Player's Navy Cut cigarettes. Given how priceless these commodities were to the British soldiers, it was an act of great generosity to share them. They were appalled by the abuses they saw perpetrated by the SS, and complained to their own guards about it. "The behavior of the English prisoners of war toward us quickly became the talk of the camp," Fritz recalled, "and the assistance they gave was of great value."

As welcome as gifts of food were, being found in possession of them would earn a whipping or days of starvation in the standing cells in the Death Block bunker: tiny, claustrophobic rooms in which it was impossible to sit down. There was one

SS man in particular to beware of. SS-Sergeant Bernhard Rakers ran the prisoner labor details in the Buna Werke as his own little kingdom, lining his pockets, sexually harassing the women workers, and dishing out savage punishments.[23] Fritz, going about with contraband food in his toolbox, was constantly at risk of bumping into him. Rakers would often search prisoners, and the discovery of any kind of contraband earned the culprit twenty-five lashes on the spot. There would be no official report—the contraband went straight into Rakers's pocket.

Fritz and the others looked for new and better ways to acquire food. It was two Hungarian Jews who came up with the ingenious idea of making and selling coats.

Jenö and Laczi Berkovits were brothers from Budapest, both skilled tailors who'd been assigned to Gustav's blackout detail.[24] One day, in a state of excitement, they approached Fritz and outlined their audacious plan. The black fabric they were using to make curtains was thick and sturdy, coated on one side with waterproofing. It would make excellent raincoats, which could be exchanged for a good price on the black market. They could be swapped for food or even be sold for cash to civilians.

Fritz pointed out the obvious problem: the curtain material was carefully stock-controlled, tallied against the number of curtains produced. Even rejects had to be handed over to Herr Ganz. Jenö and Laczi brushed this problem aside; they were positive they could siphon off a proportion of fabric. A skilled tailor could organize the usage of material so that the garments would come out of the normal percentage of wastage. With the number of curtains being produced, that would make a lot of coats. Fritz consulted his father, who agreed to give the plan a trial.

Between them the brothers managed to turn out between

four and six overcoats a day, without noticeably increasing the overall consumption of material. Meanwhile the other workers in Gustav's shop labored extra hard to keep curtain production up to speed.

The scheme had scarcely got going when it came to an abrupt halt. The brothers realized that they'd overlooked one important factor: they had no buttons, nor anything that could be used as a substitute. They asked around, and one of the Czech curtain fitters offered to bring back a supply on his next trip to Brno. With that problem solved, production resumed.

Distribution was Fritz's responsibility. He had made friends with two Polish civilian women in the insulation workshop next door, Danuta and Stepa; they smuggled the finished coats out to their labor camp and sold them to fellow workers. Others were sold to civilians in the factories. The price per coat was either two pounds of bacon or a pint of schnapps, which could be exchanged for food.

The coats, which were well made and practical, quickly became popular, which brought with it a growing danger of the SS noticing all the civilians suddenly wearing the distinctive black garments. This risk was reduced somewhat when German engineers and managers started acquiring them; these influential men now had a vested interest in turning a blind eye to the operation. Accordingly, the number of prisoners Fritz and his friends were able to help increased, and more lives were saved.

17. Resistance and Betrayal

אח

Despite all he was doing to save lives, Fritz craved a more direct form of resistance. What he really wanted to do was fight, and he was not alone.

Putting up an armed resistance against the SS was impossible without weapons and support. As things stood, the only way to achieve that would be to make contact with the Polish partisans in the Beskid Mountains. Messages could be smuggled to them fairly easily, but developing a proper relationship would require a meeting in person. Somebody would have to escape.

Word was passed to the partisans, and at the beginning of May a five-man team of escapers was chosen by the resistance leadership. First up was Karl Peller, a thirty-four-year-old Jewish butcher and one of the old Buchenwalders. Then there was Chaim Goslawski, the senior in block 48 who had looked after Fritz after his staged death. As a native of this region, if anyone could find a way to the partisans, it would be him. There was also a Jew from Berlin whose name Fritz never knew, plus two Poles known only to Fritz as "Szenek" and "Pawel," who worked in the camp kitchen.[1]

Fritz was brought into the circle by Goslawski. His role was to obtain civilian clothes for the escapers from the Canada store.

All the preparations were in place when, one morning in the predawn dark before roll call, Goslawski came to Fritz and handed him a small package, about the size of a loaf of bread.

"Give this to Karl Peller," he said softly, and melted away into the darkness.[2] Fritz secreted the package inside his uniform and rejoined his block-mates as they marched off to roll call. He had been kept out of the inner planning circle, but he guessed that the moment for the escape must be imminent.

Later that morning, on his curtain-fitting rounds, Fritz invented an excuse to visit the building site in the Buna Werke where Peller worked and slipped the package to him. At noon, Szenek and Pawel arrived in the Buna Werke with the lunch-time soup for the prisoners. Fritz noticed that Goslawski had found some pretext to accompany them. All the escapers were now inside the Buna Werke, which was far less heavily guarded than the camp.

Fritz went about his work and saw nothing more. At roll call that evening, all five men—Peller, Goslawski, Szenek, Pawel, and the Berliner—were missing. They had simply walked out of the Buna Werke wearing the civilian disguises Fritz had supplied and disappeared. While the SS launched a search, the prisoners were kept on the roll-call square under guard.

The hours ticked by. Midnight came and went, the early hours of the morning wore away, and dawn found them still standing at attention, surrounded by a chain of armed sentries. Breakfast time passed. An agitated whisper went through the ranks: the SS were not only seeking the five missing men but also an unidentified prisoner who had been seen talking to Karl Peller on the building site the previous morning.

Fritz's heart shrank in his breast; if he were identified, it would be the bunker for him this time, and the Black Wall. Despite his fear, he inwardly rejoiced. The escape had been a success.

Eventually the prisoners were ordered off to work. Away they went with empty bellies, exhausted but with spirits raised. Days went by, and despite the rumor, nobody identified Fritz as the mystery person who'd spoken to Peller. Three weeks passed with no word . . . and then, without warning, the blow fell.

The two Poles, Szenek and Pawel, along with the Berliner, were brought back to the camp, battered and disheveled. The resistance leadership learned that the three had been arrested by a police patrol in Cracow.[3] This was mystifying—Cracow was nowhere near the Beskid Mountains, almost in the opposite direction in fact. And where were Goslawski and Peller? Had they managed to join up with the partisans?

At roll call that evening, the three recaptured men were put on the *Bock* and whipped. And that, astonishingly, was the sum total of their punishment. Sometime later, when a transport of Poles was sent to Buchenwald, Szenek and Pawel were put on it.[4] The Berlin Jew remained in Monowitz.

Eventually the whole story came out. Having been too scared to speak while the two Poles were still in the camp, the Berliner revealed to a friend what had happened after the escape. The package Fritz had conveyed from Goslawski to Karl Peller had been at the root of it. It had been stuffed with cash and jewelry stolen from the Canada store, intended as payment to the partisans to secure their assistance. A rendezvous had been prearranged, but Goslawski and Peller never got there; on the first night, both were murdered by Szenek and Pawel, who took the loot for themselves. The Berliner had been too terrified to intervene.

Instead of running off with their booty, the three decided to head for the rendezvous after all. When they got there, the partisans were waiting for them. They weren't happy; they'd been told to expect five men—where were the other two? Szenek and Pawel feigned ignorance, but the partisans weren't satisfied with their excuses and evasions. They sheltered the three men for a week, but when Goslawski and Peller still didn't show up, they called off the deal. The three men were driven to Cracow and dumped. Lost and friendless, they simply wandered the streets until they were picked up by the police.

The Berliner's confession found its way to the camp senior, who passed it on to the SS administration.

A few weeks later, Szenek and Pawel reappeared in Monowitz, brought back from Buchenwald on SS orders. A gallows appeared on the roll-call square and the prisoners were ordered out on parade.

Fritz and his comrades marched into the square to find a cordon of SS troopers in front of the gallows with machine pistols leveled. The prisoners formed ranks, and in the silence that followed, Commandant Schwarz and Lieutenant Schöttl mounted the podium. "Caps off!" came the order over the loudspeakers. Fritz and eight thousand others tucked their caps under their arms. From the corner of his eye, Fritz saw the two Poles marched in. Schöttl read the sentences into the microphone: both men were condemned to death for escape and for two counts of murder.

First Szenek was led up to the gallows, then Pawel. In typical SS fashion there was no drop; a thin cord noose was put around each man's neck, and then they were hauled suddenly off their feet, legs kicking and body jerking, twitching with diminishing force. Minutes passed, and eventually they were still.[5] The commandant, having delivered an instructive lesson to his prisoners, dismissed the parade.

The whole disastrous affair weakened the Monowitz resistance. Not only had they lost Goslawski and Peller, but all the old tensions and mistrust between Poles and German Jews were revived.

It also turned the SS violently paranoid. Not long after, they claimed to have uncovered an escape plot among the roofing detail. The suspects were taken to the Gestapo bunker and subjected to horrific torture. On Commandant Schwarz's orders, three were hanged, in a repeat of the same dreadful ritual.[6] More hangings followed.

For Fritz, this was one of the most dispiriting periods in his entire time at Auschwitz. However, he had still not seen the worst.

אחים

In the late afternoon of Sunday, August 20, the first bombs fell out of a clear blue sky. One hundred and twenty-seven American bombers, flying from a base in Italy, drawing a comb of vapor trails five miles above Auschwitz, dropped 1,336 bombs, each one a quarter ton of steel and high explosive.[7] They detonated in the central and eastern end of the Buna Werke.

While the SS hid in their bunkers, the prisoners had to take their chances in the open, amid the titanic roar of explosions, with the concussions shaking their bodies. The flak batteries around the perimeter replied, thudding and hammering. Prisoners working in the factories threw themselves to the floor for protection, and rejoiced. "The bombing was really a happy day for us," one of them recalled. "We thought, they know all about us, they are making preparations to free us." Another said, "We really enjoyed the bombing . . . We wanted once to see a killed German. Then we could sleep better, after the humiliation never to be able to answer back."[8]

The bombs left the ground in and around the Buna Werke pocked with smoking craters. Most had failed to hit anything, but some buildings in the synthetic oil and aluminium plants had been torn apart, along with various sheds, workshops, and offices. Some stray bombs had landed in the camps around the factory complex, including Monowitz. Seventy-five prisoners were killed in the raid, and over 150 injured.[9]

Many Jewish prisoners were elated at seeing the SS terrorized, but some felt the opposite. The young Italian Primo Levi, who had arrived in Monowitz in February, believed that the bombing hardened the will of the SS and brought about a solidarity between them and the German civilians in the Buna Werke. Also, the bomb damage interrupted water and food supplies to the camp.[10]

The resistance was disappointed. The appearance of bombers had prompted speculation that the Allies might start parachuting in soldiers and weapons. But although American planes were seen again high overhead on a few occasions, neither bombs nor parachutes fell; they were reconnaissance flights, carefully photographing the IG Farben works and the Auschwitz complex.

What really occupied the thoughts and debates of the resistance was the relentless advance of the Red Army from the east. They had reason to fear that when the moment came the SS would carry out a mass liquidation of the whole camp, murdering all the prisoners before they could be liberated. They had done so at Majdanek.

Escape attempts continued. In October four prisoners on an outside work detail overpowered their SS guard, seizing his rifle and destroying it before making their escape.[11] Another man walked out of the camp disguised in a stolen SS uniform. He managed to get all the way to Vienna before the Nazis caught up with him, and he died in a shootout with the Gestapo.

Individual actions were inspiring, but the Jewish resistance—including Fritz—wanted more. Now that relations with the Poles had been soured, it would be impossible to hook up with the partisans. Instead, it was suggested that they try to make contact with the Red Army. In order to do that, they would need to build a relationship with the Russian POWs held in a separate enclosure in Monowitz. They could be approached via some of the Russian Jews known to the resistance. It would be hard, because there were no loyal communists or Jews among the POWs—they had all been shot immediately on capture—so there was little common ground. Nonetheless, Fritz and the others had to try. Eventually, one of the Aryanized Jews succeeded in escaping with a handful of Russians. Everyone waited anxiously for developments, and when none came they guessed he had evaded recapture.

This gave the resistance a glimmer of hope, but it was faint. Sitting in on their meetings, Fritz felt a growing impatience. His thoughts were still on fighting back when the final massacre began; hoping for Russian help felt vain and inadequate. "If we were to be slain, we should at least take a few SS men with us," he reasoned. He turned this thought over and over in his mind, but with no idea how to accomplish it, he kept it to himself.

אבא

In September the American bombers returned, aiming for the oil plant in the Buna Werke. Some went off course and dropped their bombs on Auschwitz I, where by chance they hit the SS barracks; another fell on a sewing workshop, instantly killing forty prisoners. A few hit Birkenau, slightly damaging the rail tracks near the crematoria and killing about thirty civilian workers.[12] Only slight damage was done to the oil plant, but about three hundred prisoners, who as always were barred from entering the shelters, were injured.

Some prisoners were glad to take the risk. Selections for the gas chambers took place weekly now, with sometimes up to two thousand being despatched from Monowitz at once.[13] The American bombs seemed to foreshadow liberation. How long could it be now?

"We are coming to winter again—already our sixth," Gustav wrote as the first frosts began. "But we are still here, still our old selves." News from the outside reported the Russians at a standstill near Cracow. "I keep thinking that our stay here will soon come to an end."

How long could it drag on?

בן

"I want you to get me a gun."

Fredl Wocher was taken aback. He and Fritz often met up during the day; normally Wocher would pass his friend some food or, on rare occasions when he'd been to Vienna, a letter or a package.

"Get you a what?"

"A gun. Can you do that for me?"

Wocher hesitated, but didn't ask what it was for; he didn't want to know. "I'll have to think about it," he said reluctantly. "It's dangerous."

"Remember all you've done for me," Fritz said. "This won't be any more dangerous than any of that."

Wocher wasn't convinced. A decorated German soldier smuggling guns for a Jewish prisoner? That wasn't merely dangerous, it was insane.

Despite his friend's reluctance, Fritz kept pushing. If there was a final massacre in Auschwitz—as seemed increasingly probable—he wanted to at least be able to defend himself and his father. If he could get enough guns, he might even be able to arm the whole resistance.

A few days later they met again in a quiet corner of the works site. Wocher looked excited. "Did you get it?" Fritz asked eagerly.

Wocher shook his head. "No. I've got a better idea. We should escape together, you and I."

Fritz's heart sank, but before he could object, Wocher rushed on. He had it all planned out. Once free of the camp, they would head southwest, making for the mountain country of the Austrian Tyrol. As a Bavarian, Wocher knew the region and could find them a safe sanctuary among the peasant mountain farmers. The Tyrol was right at the nexus between the two Allied fronts: American and British forces were pushing hard into northern Italy, while Patton's Third Army was driving toward the Rhine from the west. In no time, both of these advances would reach the Tyrol, and Fritz and Fredl would be liberated. "It's better than waiting here and hoping to survive," Wocher argued. Having seen the pitiless violence of the eastern front, he knew the callousness of the Red Army matched anything the SS was capable of.

Fritz was swayed by the strength of his friend's argument. But he shook his head. "It's out of the question."

"Why?"

"I'm not leaving my papa behind."

"So we take him with us."

"He's too old to survive a journey like that on foot." Actually, Fritz wasn't at all sure of that, but even if it was physically possible, he doubted his father would agree to go; there were too many people here who depended on him, and he wouldn't forsake them. There was another issue: if Fritz went without him, as Fritz's kapo Gustav might be held responsible for his escape.

"It's impossible," Fritz said. "What I need is a gun. Can you get me one?"

The German reluctantly gave in. "I'll need money," he said. "Reichsmarks won't do—it has to be American dollars or Swiss francs."

בן

The first person Fritz tried as a potential source of cash was Gustl Täuber, who worked in the Canada store. It was a haunting place, stuffy, filled with racks of coats and jackets, folded trousers, sweaters, shirts, bundles and heaps of unsorted stuff, shoes, suitcases, each with a name and address painted— a Gustav or a Franz, a Shlomo or a Paul, Frieda, Emmanuel, Otto, Chaim, Helen, Mimi, Karl, Kurt; and the surnames: Rauchmann, Klein, Rebstock, Askiew, Rosenberg, Abraham, Herzog, Engel; and over and over again: Israel and Sara. Each with a truncated address in Vienna, Berlin, Hamburg, or just a number or birth date. Every aisle between the racks and shelves was redolent with their odors, their sweat and perfumes, mothballs and leather, serge and mildew.

Gustl Täuber was an old Buchenwalder, close to Fritz's papa in age, a Jew from Silesia* born in the high days of the German Empire.[14] Fritz had never liked Täuber much; he was one of the very few who felt no bond of solidarity with his fellow prisoners, and wouldn't put himself out for anyone. But he was Fritz's best hope. They'd had a trading relationship for some time based on bonus coupons, which Täuber had used to buy vodka and (as an Aryanized Jew) visits to the brothel. Fritz knew that there was often money found in the clothing, and that Täuber pocketed whatever he could. Could he spare some? The old man shook his head. Fritz pleaded, but Täuber was immovable; he knew Fritz was connected with the resistance, and wasn't willing to put his privileges in jeopardy. Fritz was disgusted; Täuber was happy enough to get involved in risky dealings when there was a brothel visit or a bottle of vodka in it for him.

From the clothing store Fritz went to the main bathhouse. New prisoners were brought here for disinfecting and shaving, and cash and valuables successfully concealed from the Canada searchers were often taken from them here. The bathhouse

* Part of western Poland.

attendant was another old Buchenwalder, David Plaut, a former salesman from Berlin.[15] Unlike Täuber he was a decent friend. Although pickings from the bathhouse were taken by the camp kapo, Emil Worgul, Fritz reckoned Plaut, who did the actual work, must manage to sidetrack a little cash for himself. Fritz spun a yarn about needing to buy vodka with which to bribe Worgul to give some of his comrades transfers to easier labor details. It worked. Plaut went to his hiding place and came back with a little roll of American dollar bills.

Fritz met Fredl Wocher the next day and gave him the money. There followed several days of anxious waiting. Then one morning Wocher showed up at their meeting wearing an expression of mingled fear and triumph.

From under his coat he produced a pistol, a military-issue Luger. He wouldn't say how he'd obtained it, but Fritz guessed it came from one of his friends in the Luftwaffe flak batteries. He showed Fritz how it worked—how to extract the magazine and load it with bullets, how to cock it and operate the safety catch. There were a couple of boxes of ammunition with it.[16] Fritz handled it with foreboding and excitement, sensing the lethal power in his palm.

Now came the problem of getting it back to camp. Contraband food was one thing; firearms were in a different league. Retreating to a hiding place, Fritz dropped his trousers and tied the Luger to his thigh. The ammunition went in his pockets. That evening he marched back to camp tingling.

After roll call he went straight to the hospital and found Stefan Heymann. Beckoning him to follow, he led his friend behind a stack of dirty laundry and showed him the Luger.

Stefan was horrified. "Are you crazy? Get rid of that thing! If you get caught with that it won't just be you they kill—you're putting our whole operation at risk."

Fritz was hurt. "You brought me up to be like this," he said indignantly. "You always taught me that I had to fight for my life."

Stefan had no answer to that. Over the next few days they talked again and again; Fritz explained his thinking—the ferocity of the battle that might take place here, the notorious brutality of the Russians, the likelihood of the SS massacring the prisoners—and gradually wore Stefan down. "I'm sure I can get more guns if I have the money," he offered.

Stefan thought it over. "Very well," he said at last. "I'll do what I can. But the whole thing must be properly organized. No more going it alone."

He managed to scrape together two hundred dollars, which Fritz took to Fredl Wocher. Another period of waiting followed, then one day Wocher led Fritz to a discreet spot in the factory and showed him where he had hidden another Luger and two MP 40 machine pistols—the distinctive submachine guns used by German soldiers everywhere. There were several boxes of ammunition for all three weapons.

This would be a much bigger challenge to smuggle into the camp. Fritz planned it carefully; it would take several trips. Obtaining one of the huge canisters used to bring soup for the prisoners' midday meal, he built a false bottom into it, beneath which he concealed the ammunition. The Luger he strapped to his thigh again, but the machine pistols were a different matter. Having been tutored in their use and maintenance by Wocher, he dismantled one and tied as many of the parts as he could to his bare torso.

With winter deepening and the nights drawing in, it was dark at the end of his shift, so there was little chance of the guards noticing his unusually bulky shape. All the same, it was stomach churning, as was standing through hours of roll call with the heavy parts strapped to him.

The moment it ended, Fritz walked swiftly to the hospital laundry, where Jule Meixner was waiting for him. Fritz hurriedly stripped off his uniform, untied the gun components, and passed them to Jule, who hid them. For security, Fritz wasn't told where—on the principle that you can't give up a secret under torture if you don't know it in the first place. Over the next several days, he repeated the dangerous operation until all three guns and their ammunition were inside the camp.

Fritz felt pleased with himself; by bringing the Luger into the camp, he had forced Stefan's hand. The resistance would never have done it without him. Now, if a repeat of Majdanek happened here, they'd be able to extract some blood from the SS in return.

אבא

Through December, Gustav's workshop went on turning out blackout curtains and coats in parallel. With no direct involvement in the resistance, he had no idea of the dangerous venture Fritz had embarked on. Gustav was looking forward to Christmas, when Wocher would be making another of his trips to Vienna.

One Monday afternoon, the workshop was running at its usual full tilt when suddenly, over the soft snickering of sewing machines, they heard the rising moan of the air-raid sirens.[17] Within seconds, doors began slamming, feet running, voices shouting. The SS and civilians were making for the shelters. Gustav's staff looked at him. He gave them permission to run off to whatever makeshift hiding places they wished. Gustav stayed where he was. Hiding places would be of little use if a bomb fell close.

After a few minutes, with the last panicked footsteps dying away, the drone of planes and the thumping of the flak guns began. The noise rose in a crescendo, and with it came the first earth-shattering concussions of bombs. Gustav lay flat; this was no new terror for him: he'd spent months under bombardment in the trenches, and had learned to sit and wait for it to either pass or for a random bomb to find him and send him to oblivion. It was both useless and dangerous to panic. His great fear was for Fritz, who was out on fitting work. Gustav knew his son had a hiding place among the buildings where he would at least be sheltered from flying debris.

Again the bombers were aiming for the synthetic oil plant, but a lot of the explosions seemed to scatter randomly—some far away, some uncomfortably close. Suddenly, the floor beneath Gustav was rocked by a titanic detonation. Windows shattered, and there was a cacophony of tearing metal and masonry. Gustav covered his head and sat tight. The shuddering died away. Dust floated in the air, and beyond the bubble of silence immediately around him Gustav could hear distant screams and yelling, the pounding of the guns stuttering to a halt, and the drone of the bombers receding. The all-clear began to howl.

Climbing to his feet, Gustav found the workshop in disarray: sewing machines shaken loose and toppled from their benches, chairs knocked over, dust everywhere, shards of glass from the broken windows. The men and women who'd stayed with him stood up, coughing and blinking.

As soon as he was satisfied that nobody was hurt, Gustav's first thought was for Fritz. He went outside into a chaos of smoke and flame. Some buildings had been destroyed, and dead prisoners lay scattered in the open and among the rubble. Injured men and women were being assisted by their comrades.[18]

There was no sign of Fritz. Gustav hurried through the smoke, heading for his son's hiding place, consumed by a rising sense of foreboding. Turning the corner, he reached the place. It was gone. Instead just a hill of broken, tumbled rubble and twisted metal. Gustav stared in shock and disbelief.

After a while he began wandering back in a daze of grief. His Fritzl—his pride and joy, his dear, sweet, loyal Fritzl—was gone.

Civilians and SS men were emerging from their shelters. Hardly any had stayed at their posts. The fences were down in a few places, and several prisoners had escaped. Gustav stood a moment staring as the SS tried to restore order. He was about to turn away when he saw two figures in stripes walking toward him through the smoke, one carrying a large toolbox and moving with a familiar gait. Gustav could hardly believe his eyes. He ran and threw his arms around Fritz. "My boy, my Fritzl, you're alive!" he sobbed, kissing the bemused boy's face and hugging him, repeating over and over, "You're alive! My boy! It's a miracle!"

He took Fritz by the arm and led him to the smoking remains of his hiding place. "It's a miracle," he kept repeating. Gustav's faith in their good luck and fortitude, which had helped keep him alive for so long, was again vindicated.

אב ובן

Another air raid hit the Buna Werke on the day after Christmas. The Americans had fixed on it as a prime target, and were determined to annihilate it. But each time they only succeeded in flattening a few buildings, wounding a handful of Nazis, killing or injuring hundreds of prisoners, and reducing productivity. Droves of slaves cleared the rubble, repairing and rebuilding. They sabotaged what they could, and worked as slowly as they dared, and between them and the bombs they ensured that the Buna Werke would never produce any rubber, and its other plants would never reach full capacity.

On January 2, 1945, Fredl Wocher returned from Vienna with letters and packages from Olly Steyskal and Karl

Novacek. "We get the greatest joy from knowing that we still have good friends at home," Gustav wrote in his diary.

Not only that, he and Fritz had the very best of friends in Fredl Wocher himself. He had proved himself countless times, and in so many ways. With the Red Army now positioned just the other side of Cracow, Fritz tried to persuade him to disappear before the Russians reached Auschwitz and discovered what had been happening here.

Wocher didn't see the need. "My conscience is clear," he said. "More than clear. And I'm just a civilian, a worker; nothing will happen to me."

Fritz wasn't convinced. He reminded Wocher of the hatred the Russians felt for all Germans—which Wocher knew all too well from his service at the front. Also, there were thousands of Russian prisoners in Auschwitz who would be thirsty for revenge the moment they got the chance. Wocher couldn't depend on being spared once the wave of vengeance swept through the camps. But he was stubborn; he'd never run away before, and he wasn't about to start now.

It was clear to Fritz that the end might come any day. His preparations had been in process for two months. Thanks to him, the resistance had weapons. At the same time, Fritz had taken the extra precaution of equipping himself and his father for escape. Having dismissed the idea of fleeing to the Tyrol, he had to accept that fighting might not be an option either. Therefore, on Fritz's initiative, he and his father had been dodging the weekly headshaving and letting their hair grow to a normal length. Roll call was the only time prisoners routinely took their caps off in front of the SS, and in the winter months that always happened during the hours of darkness. Fritz had also acquired a cache of civilian clothes from David Plaut at the bathhouse, which he hid in a toolshed in the camp. There were enough jackets and trousers for himself, his papa, and a few of their closest comrades.

On January 12 the Red Army launched its long-anticipated winter offensive in Poland—a colossal, well-planned assault along the front line, involving three armies of two and a quarter million men. This was the final push, designed to drive the Germans back into their Fatherland. The Wehrmacht and the Waffen-SS, outnumbered more than four to one, fell back under the onslaught, holding out in a handful of fortified Polish cities. Frustratingly, the sector of the front near Cracow moved slower than most. Each day the prisoners in Auschwitz heard the distant thump of Russian guns, like a clock ticking away the moments to deliverance.

On January 14 Alfred Wocher said a last goodbye to Gustav and Fritz. He had been drafted into the Volkssturm, a hastily organized army of old men, underage boys, and disabled veterans, tasked with the last-ditch defense of the Reich. So he would not be found by the Russians at Auschwitz after all. He was happy to do this final duty for his Fatherland. Whatever he felt about its crimes, it was Germany after all, his home, a land full of women and children, and the Russians would tear it apart without mercy if they were permitted.

With winter deepening, the weather was deteriorating. There was thick snow, and on Monday, January 15, the day after Wocher's departure, Auschwitz awoke to thick fog. The prisoners in Monowitz were kept standing at roll call for several hours until the fog thinned enough for the SS to feel safe marching them to work.[19]

In the factories, work went on at full pace. The previous night, an American plane had flown over, illuminating the whole area with parachute flares and taking photographs. Photos taken twenty-four hours before had shown nearly a thousand bomb craters in the factory complex and forty-four wrecked buildings, but the nighttime images revealed that repairs were well in hand, and that the synthetic fuel plant—the most important of all—was virtually untouched.[20]

On Wednesday the prisoners were held back at roll call again. They remained on standby throughout the morning, and in the afternoon they were marched to the factories. But after only two and a half hours' work they were marched back to camp again.

The SS were getting jittery. Each morning, the rumble of artillery was a little less distant. By evening on the seventeenth, the sound was nearer still, and the Auschwitz commandant, Major Richard Baer, at last gave the order to begin evacuating the camps. Invalids were to be left behind, and any prisoners who resisted, delayed, or escaped were to be shot immediately.[21] The leader of the Auschwitz I resistance alerted his partisan contacts in Cracow: "We are experiencing the evacuation. Chaos. Panic among the drunken SS."[22]

That evening, all the patients in the prisoner hospital in Monowitz were examined by the doctors; those who were well enough to march were struck off the patient list and herded back to their blocks. The rest—numbering over eight hundred—were left to the care of nineteen volunteer medical staff.[23]

The following day, Thursday, January 18, all eight thousand prisoners in Monowitz were kept standing on the roll-call square all day in bone-aching cold. Fritz and Gustav, aware that the end was imminent, had put on their civilian clothes under their uniforms, ready to make a break for it the moment they got the opportunity. At least with their extra layers they were slightly less cold than their comrades. Dusk began to gather.

Finally, at 4:30 p.m., the SS guards began ordering the prisoners into columns. With their limbs numb and joints seizing up, they were arranged like an army division into company-size units of about one hundred, further grouped into battalion-size units of about a thousand, which in turn formed three larger units, each containing up to three thousand. SS officers, Blockführers and guards took command of each unit.[24] Anticipating trouble, every SS man had his rifle, pistol, or machine pistol ready in his hands, with the safety catch off. Fritz thought regretfully about his guns, concealed somewhere in the hospital laundry. It was impossible to get anywhere near them now.

Disturbingly, the infamous SS-Sergeant Otto Moll was on hand. He wasn't part of the Monowitz guard battalion—he'd been director of the Birkenau gas chambers—yet here he was, walking among the waiting columns as they were issued with their marching rations, dishing out abuse while they got their bread, margarine, and jam. Moll, a stocky little man with a bull neck and a head as wide as it was high and the blood of tens of thousands on his hands, made a deeply unsettling presence in these circumstances. He stopped beside Gustav, drawn by something about his appearance, looked him up and down, then gave him two hard slaps across the face, left and right. Gustav staggered and recovered. Moll moved on without a word.[25]

At last the order was given and the columns began to move. Tired already from standing all day, they marched off the square, five abreast, wheeling left onto the camp street. Passing the barrack blocks, the kitchens, the little empty building where the camp orchestra had lived, the massed prisoners passed out through the open gateway for the last time.

They were leaving a place which for a few had been home for more than two years. The old survivors like Gustav and Fritz—especially Fritz—had helped build it from grassy fields; their comrades' blood had gone into its construction, and pain and terror had been the unrelenting life of the place ever since. Yet it was home nonetheless, by the simple virtue of the animal instinct to belong, to attach oneself to the place where one ate and slept and shat; however much one hated it, it was where friends were, and where every stone and timber was familiar.

As for where they were going, they didn't know. Away from the Russians, that was all they knew. All the Monowitz subcamps were on the move—over 35,000 men and women[26] taking to the snow-lined roads leading west from the town called Oświęcim.

PART IV

Survival

18. Death Train

Fritz sat on the ground close beside his papa, shivering convulsively. Around them sat their friends. It was early morning, and the cold was beyond imagining. They had no shelter, no food, no fires: just one another. They were almost dead from exhaustion and exposure. Some would never get to their feet again when this rest stop was over.

For the first few miles after leaving Monowitz, Fritz and Gustav and the other reasonably healthy prisoners had helped their weaker comrades along. Anyone who lagged behind was beaten by the SS with a rifle butt and driven onward. If a person fell down in the middle of the pack, the semiconscious marchers following behind would trample over him. Fritz and the others did what they could, but comradeship only stretched so far. They were scarcely past Oświęcim when they ran out of strength and had to leave the weakest to fare as best they could. They hugged their jackets tight about them and closed their ears to the sporadic gunshots from the rear of the column as stragglers were murdered.

To Fritz and Gustav it was like a repeat of the forced march, so many years ago, along the Blood Road to Buchenwald. But this was infinitely, inconceivably worse. They kept close for protection, father and son, heads down, one foot in front of the other on the compacted snow and ice, numbed in mind and spirit, hour after hour through darkness swirling with white flakes. Close by Fritz a Blockführer marched, pistol in hand;

Fritz could sense the man's terror of the pursuing Russians, and the violence in him.

By Gustav's reckoning they had trudged twenty-five miles when they reached the outskirts of a town in the dawn light. The column was directed off the road into an abandoned brickworks. The SS guards needed a rest almost as much as their charges. Finding what shelter they could among the stacks of bricks, the prisoners sat close for warmth. Fritz and his papa stayed awake, despite their consuming weariness, guessing that anyone who slept would never wake again. Talking with some comrades who had been in different parts of the column, they discovered that several Poles—including three of Fritz's friends—had escaped.

"We should do that," Gustav said to Fritz. "We should make a run for it. I speak Polish; we'd have no trouble finding our way. We could find the partisans or just head for home."

For all his preparations, all his determination to resist, Fritz's heart quailed at the thought. There was one very big problem: he spoke no Polish. If they were separated, he'd be sunk.

"We should wait until we reach German soil, Papa," he said. "Then we'll both speak the language."

His father shook his head. "It's a long way to Germany." He looked around at their exhausted comrades. "Who knows if we'll ever reach it? Even assuming the SS intend us to survive that far."

Their discussion was cut short by the order to move. When they hauled themselves to their feet, some men who had fallen asleep remained where they were. Hypothermia had taken them, and their bodies were already beginning to freeze. Others were still alive but too weak to stand; the SS went among them, kicking and chivying, shooting any who couldn't be roused.[1]

The column trudged on. Behind them stretched a nightmare of trampled snow and scattered corpses, leading all the way back to Auschwitz, where the last evacuations were still in progress. Jews too weak to evacuate were being forced to burn the stacks of corpses around the gas chambers. The crematoria were dynamited and SS clerks burned records. Some pilfered from the Canada stores, where the incriminating mountains of loot were also being put to the torch. In the end the sheer weight of the crimes committed here would defy all efforts to erase the evidence.

That evening, the column reached the town of Gleiwitz,* where there were several subcamps belonging to the Auschwitz system. The Monowitzers were herded into an abandoned enclosure that had been built for only a thousand inmates. The prisoners had been evacuated the previous day.[2] The Monowitzers were given nothing to eat, but were thankful at least for shelter in which they could sleep.

Two days and nights they remained in Gleiwitz while the SS organized the next stage of the journey. Unlike most of the poor souls marching from Auschwitz, the men of Monowitz would be going by train.

Rousted from their huts, they were herded to the city freight yard, where their transports were waiting. Instead of the usual closed freight cars, the four long trains were made up of open-top wagons, normally used for carrying coal and gravel. Rations were doled out—half a loaf of bread each, with a piece of sausage—then the loading began. Fritz and Gustav climbed into a wagon along with more than 130 other men, clambering up the sheer sides and dropping down with a clang on the steel floor, which echoed less with each pair of feet until the last few had to squeeze in between the rest.

Every other wagon had a brake house—a little hut raised above the level of the wagons. In each one an SS guard was

* Now Gliwice, Poland.

posted, armed with a rifle or machine pistol. "Anyone putting their heads above the sides will be shot," warned the Blockführer in charge of loading.

The train began to vibrate. Steam and smoke from the locomotive made a thick fog in the icy air. At last, with the clang and bang of couplings and shriek of wheels, the train moved, dragging its load of four thousand souls.[3] As it built up speed, the wind, chilled to four degrees below zero, roared across the open wagons.

אבא

The Holocaust was a crime made of journeys, criss-crossing Europe to the accompaniment of a tuneless score of protesting machinery. Wheels hissed on the rails; couplings groaned and jolted: the hissing-squealing-clanking-banging of steel-wheeled boxes on metal rails was a never-ending nightmare music.

Gustav's body rocked from side to side with the motion of the train. He sat with his knees drawn up to his chest, with Fritz close beside him, hugged up against the terrible cold.

After leaving Gleiwitz, this train had diverged from the other three, heading south while they went west. The next morning it stopped to take on hundreds more prisoners evacuated from the Charlottengrube subcamp[4] before crossing into Czechoslovakia. Despite the Blockführer's warning, Gustav peeped over the side from time to time, gauging the progress of their journey, noting the towns through which they passed. The train never stopped, but it went painfully slowly, and took two freezing nights and a day to cross Czechoslovakia.

They'd been told that they were being taken to Mauthausen concentration camp. The thought was simultaneously thrilling and terrifying for the Austrians; Mauthausen's reputation for violence was dire. But it was in Austria, in the beautiful hill country near Linz. *Austria!* Soon Gustav and Fritz would be on their home soil for the first time in over five years.

And there they would surely die. In Mauthausen they would have none of the support system they'd built up in Auschwitz, and would be subjected to an even harsher regime.

That was assuming they made it that far. Even as Gustav turned these thoughts over in his mind there was a stir among his fellow prisoners. Another had passed away. Weakness, sickness, and hypothermia had been killing them off steadily. A friend of the dead man stripped the jacket and trousers off the body and put them on over his own clothes in an attempt to keep the cold out. The body was passed across the wagon and stacked in the corner with the other corpses, all stripped to their underwear and frozen solid. That corner also served as the latrine, and even in the cold the stench was abominable.

The deaths at least created more room to sit. Gustav looked around at the gaunt faces, the deep shadows beneath the eyes, the cheekbones whittled to ridges by starvation. Some had managed to eke out their ration, and as the fourth day of the journey passed, they nibbled their last crusts. Gustav and Fritz had none left. Gustav could already feel his strength slipping away; a slow ebb tide eroding his will. Only one thought was on his mind now: escape.

"We have to go soon," he said quietly to Fritz. "Otherwise it'll be too late." If they could slip over the side during the night, they might not be noticed by the guards. Soon they'd be in Austria, and language would be no problem. They could make their way to Vienna in their disguises and find a hiding place. "Olly or Lintschi will take care of us."

"All right, Papa," said Fritz.

That night they tested the watchfulness of the guards. With help from a couple of friends, they lifted a corpse from the stack, heaved it up to the rim of the sidewall, and pushed it over. As it went flailing away into the darkness, they waited for a yell from the brake house and a burst of gunfire . . . but nothing came. This would be easy. All they had to do was wait until they crossed into Austria.

By morning, the train reached Lundenburg,* just a few miles from the Austrian border. There, frustratingly, it halted. Hour followed hour, and nothing happened. A peek over the side showed that the whole train was surrounded by SS. It was dusk when they finally started moving again and the Czech countryside gave way to Austria. Now was their time. With each passing mile the situation in the wagon was getting worse, descending into savagery. Some of their comrades had reached the point where they would strangle a friend for a mouthful of bread. Through cold, hunger, and murder, the corpses were piling up in the corner at the rate of eight to ten every day.

Fritz nudged his father. "Papa! Wake up! It's time to go."

Gustav drifted awake and tried to rise. He couldn't get up; his frozen muscles were too weak. He looked at Fritz's eager face. "I can't do it," he said.

"You have to, Papa. We have to leave while we can."

But nothing Fritz said could raise him. "You have to go alone," said Gustav feebly. "Leave me and go."

Fritz was appalled at the very idea. *If you want to go on living, you have to forget your father.* That was what Robert Siewert had said to him that day in Buchenwald. It had been impossible then, and it was impossible now.

"You have to go," Gustav insisted. "I can't make it—I'm old, my strength is gone. Leave now—*please*."

"No, Papa. I won't." He sat down again and wrapped his arms around his father.

* Now Břeclav, Czech Republic.

When dawn came, they found themselves in familiar snow-laden countryside near Vienna. The train steamed along the north bank of the Danube, and in broad daylight rolled into the northern suburbs and across the river into Leopoldstadt. They scarcely dared peek out as their home passed by, heartbreakingly close. They rolled by the west end of the Prater, and then the train was rumbling over the Danube Canal, through the western suburbs and back out into open countryside.

In the late morning they passed through the town of St. Pölten, and in the afternoon reached Amstetten, where the train halted. They were now barely twenty-five miles from Mauthausen.

When darkness fell, the journey resumed.

Gustav again pleaded with Fritz to escape. "You must, before it's too late. Please go, Fritzl. *Please*."

Fritz gave way. The pain of it would never leave him: "After five years of shared destiny, that I should now sever myself from my father," he recalled in anguish.

The train had reached its maximum speed. Fritz stood and peeled off the hated striped uniform with its *Judenstern* and camp number and flung off his cap. He embraced his papa one last time and kissed him, then with help from a friend he climbed the sidewall.

The full force of the subzero wind stabbed his body like spears. The train shook and thundered. He peered anxiously toward the brake house. The moon was brighter now than it had been when they'd tested the guards' alertness: two days from the full, an eerie glow illuminating the white landscape and the trees flying past.[5]

Gustav felt a last squeeze of his hand; then Fritz launched himself into the air. In an instant he was gone.

Sitting alone on the floor of the wagon, by the light of the moon Gustav wrote in his diary: "The Lord God protect my boy. I cannot go, I am too weak. He wasn't shot at. I hope my boy will win through and find shelter with our dear ones."

The train sped on, hammering and clanking, as if the locomotive itself was desperate for this dreadful journey to be over. It passed through Linz in darkness, crossed the Danube, and doubled back east toward the small town of Mauthausen.

בן

Fritz tumbled through the air, all sense of space and direction lost for a fleeting moment. The ground hit him violently, jarring his bones and knocking the wind out of him. He rolled over and over through the thick snow and came to rest with the train wheels clattering past his face, not daring to move a muscle.

The last wagon rushed by and faded into the distance, leaving him alone in the silence under the vault of stars. He looked around. He was lying in a thick drift, which had cushioned his fall. Despite the pains in his limbs, he hadn't broken anything. Shaking himself down, he started walking back along the tracks toward Amstetten.[6]

Nearing the town, Fritz's nerve failed him. He wasn't ready to face entering a town, even late in the evening. Slithering down the embankment, he struck out across an open field. It was hard going, with snow up to his hips, but eventually he came to a narrow backstreet on the edge of town. It was deserted. Warily, he followed it.

He managed to skirt around the north of the little town without meeting anyone, and was soon on a country road winding eastward, parallel with the railway line. He passed through several small villages and hamlets, gradually working his way back in the direction of St. Pölten. It was slow going on the slippery roads, and his strength was faltering.

After several hours he reached the little town of Blindenmarkt, where the road converged with the railway. The train had passed through this place the previous day. There was a small station where the passenger trains between Linz and Vienna stopped. He was tired, and in his pocket he had some Reichsmarks—his little stock of emergency cash scavenged in Monowitz. Should he risk it?

On an impulse Fritz turned off the main road and walked to the station. It was still dark, so he found an empty cattle wagon standing on the tracks and crawled inside. It was too cold to sleep, but at least he was out of the wind.

Toward dawn, lights came on in the station building. Fritz waited a few minutes, then summoned his courage and dropped down from the wagon.

The building was quiet, with just a solitary clerk behind the ticket window. Fritz hesitated; he wasn't certain what the proper procedure was nowadays. Would he be asked to show his papers? He approached the window and, as casually as he could, asked for a ticket to Vienna. The clerk, who wasn't accustomed to people traveling this early, regarded him with some surprise (and suspicion, it seemed to Fritz). But he took Fritz's money without a word and gave him his ticket.

Fritz went into the deserted waiting room and sat down. After a few minutes, the clerk came in and lit the stove. Fritz moved closer to it—the first warmth he'd felt since leaving Monowitz. He was cold to the marrow, and the sensation of life and heat flowing into his body was both heavenly and torturous, filling the deadened nerves with pins and needles and awakening the aches of his journey.

Drowsy with fatigue, he had no idea how long he'd been sitting there when the Vienna train finally huffed to a stop outside the window. Fritz went out to the platform—still the only person there—and got into one of the third-class carriages.

Closing the door behind him, he saw with a jolt of horror that the carriage was full of German soldiers. Not a single civilian—just a crowd of field-gray Wehrmacht uniforms. Luckily, they were too busy talking, smoking, playing cards, and dozing to take any notice of him. It was too late to get off again, so he found a space and sat down.

As the train moved off, Fritz glanced surreptitiously around him. He felt like a foreigner in his own homeland, with no idea of laws or protocol, and little notion how to behave like an ordinary civilian. The soldiers scarcely glanced at him. Listening to their chatter, he guessed they were returning from the front on leave.

After a couple of hours and a few more stops (at which nobody else got on), the train reached St. Pölten, where it halted. Two German soldiers came aboard, both wearing the distinctive steel gorgets of the Feldgendarmerie—the Wehrmacht military police.

They made their way along the aisle, demanding to see passes. The soldiers sitting near Fritz took their identity cards and passes from their breast pockets. Fritz took out his ticket, which was all he had. The soldiers bundled their documents together and handed them all at once to the nearest policeman; Fritz, seizing his opportunity, slipped his ticket in among them.

The policeman glanced in turn at each of the soldiers and handed back their documents. Then he came to the solitary rail ticket and frowned. He looked at Fritz, and gestured impatiently. "Papers, please," he said.

Heart pounding, Fritz made a show of riffling through his pockets. He shrugged helplessly. "I've lost them."

The policeman's frown deepened. "All right. You'd better come with us."

Fritz's heart sank, but he knew better than to argue. He got up and followed the Feldgendarmes off the train.

"Please, I need to get to Vienna," he said as they led him away.

"We can't let you go any farther till we establish your identity."

They led him out of the station to a nearby Wehrmacht outpost. He was questioned sternly, although not aggressively, by a sergeant.

"Why did you board that train?"

"I need to get to Vienna," said Fritz.

"But why that train in particular? You must have known it was a front-line special. There'd have been a regular train not long after."

"I—I didn't know."

"A young fellow in civvies with no papers boarding a troop train. That's not normal, is it? What's your name, lad?"

"Kleinmann. Fritz Kleinmann." He saw no point in lying. It was a perfectly acceptable German name, and hardly unique.

"Why don't you have papers?"

"I must have lost them."

"Home address?"

On the spur of the moment, Fritz gave a fictional address in a town near Weimar. The sergeant wrote it down.

"Stay there," he said, and left the room.

He was gone a long time, and when he reappeared he was accompanied by a superior. "We checked the address you gave. It doesn't exist. Now, where do you really live?"

"I'm sorry," said Fritz. "My memory plays tricks on me." He gave them a different address.

They went away, and again found it was false. By now, Fritz was desperately playing for time. The Feldgendarmes went through the charade once more, disposing of a third fake address before finally losing their patience.

Two guards were summoned. "Take Herr Kleinmann to the barracks," the sergeant ordered. "The security section."

He was put in a vehicle and driven through the streets to a small army barrack complex, where he was taken to a jail-like building with an office and cells.

An officer looked over the note from the Feldgendarmerie and asked Fritz to identify himself properly. "If you lie to me, I will lock you up."

What else could Fritz do? He gave a fourth imaginary address. It was checked, and he was placed formally under arrest. The officer was calm and quiet; he didn't yell or rage or threaten torture; he simply directed his men to lock Herr Kleinmann in a cell. "Perhaps the truth will come to you in there," he said ominously.

The cell was large, with three prisoners—all soldiers—already in residence, awaiting court-martial for minor offenses. They regarded him curiously, and Fritz fell into desultory conversation, explaining simply that he was a civilian who'd lost his papers and was waiting for verification.

It was pleasantly warm in the cell. It had a bed for each man, a table and chairs, and a basin and toilet in the corner. Fritz hadn't been in such comfortable accommodation in years. When an orderly brought their evening meals—the first hot food Fritz had had in nearly a week, and his first full meal for as long as he could remember—he had to force himself to eat in normal mouthfuls rather than gobbling it down like a ravenous dog.

After dinner, when Fritz turned back the blanket on his bed, he could hardly believe his eyes—there were sheets underneath. *Sheets!* What kind of a cell was this? Easing his exhausted body into bed was little short of heaven, and he slept soundly and blissfully through the night.

The next morning was, if possible, even better. The orderly brought breakfast, and simple as it was, it made Fritz's head spin. There was *real* hot coffee, bread, margarine, sausage, and plenty of it. While his cellmates chatted idly, Fritz kept his head down and concentrated on filling his stomach.

Eventually he was brought before the officer again, who demanded to know who he really was. As the questioning went on, Fritz began to realize that the officer was working on the theory that he was an army deserter. It made perfect sense. His age, appearance, and accent were all consistent with it, as were the circumstances of his apprehension. Believing that he'd caught his prisoner in one deception, the officer didn't think to look for another enormous one—that this young man with the chiseled features, civilian dress, and Viennese accent might actually be a Jew on the run from the SS.

Fritz refused to answer any further questions, and was put back in the cell. He felt contented in there: safe, warm, and well fed. Lunch consisted of a simple but very good stew and a piece of bread. Yes, this was enough for contentment.

And yet, despite these luxuries, the part of Fritz's mind that had kept him alive in the camps was fully aware of the danger he was in. Sooner or later this officer would find out the truth. As the day wore on, Fritz groped around for a solution. After dinner that evening, while his cellmates were busy talking, he surreptitiously pilfered a stick of shaving soap from one of them and ate it. By the next morning, he was violently ill: hot, sweating, and with terrible diarrhea.[7] His cellmates called the guard, and Fritz was carried out.

They took him to a military hospital. During the examination—which discovered nothing more serious than stomach cramps and a raised temperature—he had his wits about him sufficiently to keep his Auschwitz tattoo hidden. He was put in a side ward by himself and kept under observation.

It was even better than the cell: crisp white bed linen, female nurses bringing him tea and medication. After a while he was able to eat, although the diarrhea persisted. A small price to pay for the postponement of his interrogation. A doctor who visited him on the third day mentioned that there was a sentry outside the door with a machine pistol, so he'd better not be thinking of making a run for it.

Eventually the fever passed and the diarrhea cleared up. Fritz was immediately returned to the security section. He was met by the officer, whose patience was wearing thin. "It's time for this case to be closed," he said. "If you don't confess, I shall hand you over to the Gestapo."

He apparently expected that terrifying threat to break the prisoner, but Fritz said nothing. Seething with frustration, the officer ordered him back to his cell. "Two more days," he promised, "and then I'm done with you!"

There followed two days of exquisite comfort, and then Fritz was brought back to the interrogation room.

"I have guessed who you are now," said the officer, to Fritz's alarm. "You're no deserter at all. I believe you're an enemy agent, on a mission for the British. You've been dropped by parachute to engage in covert operations." Having delivered this astonishing judgment, the officer stated flatly: "You will be treated as a spy."

Fritz was appalled; this was worse than if he'd been identified as a concentration camp escapee. He denied the accusation strenuously, but the officer refused to listen. In his mind, only an enemy agent would be sneaking about in the way Fritz had been, associating with German troops. And only a trained spy would be able to resist interrogation for so long. No deserter could do that.

Despite his denials, Fritz was force-marched back to his cell. Suddenly it didn't feel quite so congenial. Should he confess?

No: he'd be returned to the SS and executed. But the outcome would be the same if they believed him to be a spy. On the other hand, even if he confessed, would they believe him now? The officer's notion of him as a German-Austrian émigré was so fixed, and he seemed so impressed with himself for having nailed a British spy, that even if he saw the tattoo he might believe it was part of Fritz's disguise.

The next day, Fritz was taken before the officer once more. Three armed soldiers were standing by. "I'm through with your denials," the officer announced, "and I'm washing my hands of you. You are going to Mauthausen. Let the SS deal with you."

19. Mauthausen

בֵּן

Fritz felt the pinch of steel round his wrists as the handcuffs snapped shut. "If you make any attempt to escape," said the officer, "you will be shot immediately."

His three-man escort—an NCO and two privates—marched him to the railway station, where they boarded a train for Linz. For the third time, Fritz traveled the familiar route: St. Pölten to Blindenmarkt to Amstetten, at some point passing the spot where he had made his leap, unidentifiable now in daylight with the snow melting. How vivid it all was in his memory. But no more vivid than his pleasant interlude in St. Pölten; like a blissful holiday, he would always remember it as lasting little more than a week, when in fact it had been closer to three.[1] Three weeks of eating well, resting in safety, and having his health restored.

At Linz they changed to a local train for the short journey to Mauthausen, a pleasant little town nestling in a bend of the Danube beneath rolling green hills checkered with fields and woods. Fritz was marched through the town two paces ahead of his guards, who kept their rifles trained on his back. The locals, accustomed to living in the shadow of the camp in the hills above the town, paid them no heed.

A winding road led up the valley. When the place came in sight, it was like no concentration camp Fritz had ever seen—more like a fortress, with high, thick stone walls topped by walkways and studded with gun emplacements. There was an angle in the wall, in which there stood a massive stone gate-house flanked at one corner by a squat round tower and at the other by an enormous square turret four stories high. Some-where within those walls were Fritz's father and friends. Or so he hoped. One could only imagine how harsh the selections would be in such a camp. But Fritz had faith in his father's strength; deep down he was certain they would be reunited—much sooner than they had expected. Fritz would certainly have a story to tell.

Instead of taking him through this imposing gate, his guards turned and marched him along the road parallel to the outer wall, past a fruit garden. At the corner, the road swung sharply right, the ground on one side falling steeply away to a sheer drop, like a vast gorge, lined with jagged cliff faces.

Fritz was looking down into the place that gave Mauthausen its evil name: the granite quarry. Larger and many times deeper than the limestone quarry at Buchenwald, its bottom was a hive teeming with slaves and echoing to the tinkling clangor of picks and chisels on stone. On the far side was a broad, steep staircase cut into the rock, curving upward in one enormous flight of 186 steps from the bottom of the pit to the rim. Up it, hundreds of prisoners were climbing, each carrying a square block of granite on his back. They called it the Stair of Death, and it was the symbol of all that was hideous about Mauthausen.

The granite extracted here was destined for Hitler's monu-mental building projects, a grandiose vision requiring stone in colossal quantities. Thousands of prisoners had died extracting it. The Stair of Death was the epitome of SS thinking—why install a more efficient mechanical conveyor when criminal and Jewish labor was so cheap and the process so satisfyingly punishing? Injuries and fatalities were constant—the slightest misstep on the staircase would send a man and his granite block tumbling among the others, setting them off like dominoes, breaking limbs and crushing bodies.

Following the road along the edge of the quarry, Fritz and his guards came to a compound of low barrack huts. Here the Wehrmacht guards handed him over to the SS and departed.

Fritz had been expecting an interrogation and a beating, but received neither. They still weren't sure what to make of him. An SS sergeant marched him to the main gatehouse, another titanic construction of granite, with two towers crowned by lookout posts decked with floodlights and machine guns. This was the main entrance to the prisoners' part of the camp (the gatehouse he'd seen at the front led to the SS garages).

Passing into the fortress, Fritz found himself in a surprisingly small and ordinary interior; it was more compact than Monowitz and filled with rows of similarly basic wooden barracks arranged on either side of a narrow roll-call ground. The sergeant disappeared into the gatehouse, ordering Fritz to wait by the wall.

A few prisoners were hanging around there. One came over and studied Fritz's civilian clothes. "Who are you?" he asked. "What're you here for?"

"My name's Fritz Kleinmann. I'm from Vienna."

The man nodded and walked off. A few moments later he came back accompanied by another prisoner, who had an air of authority, clearly some kind of functionary.

"You're from Vienna," he said. "Me too. Been here for years." He studied Fritz. "This place is pretty bad, but the one thing you really don't want to be here is a Jew. Jews last no time at all." With that, he walked off.

Eventually the sergeant emerged from the gatehouse and, to Fritz's surprise, asked whether he had an Auschwitz tattoo. There had been a few transports in from Auschwitz lately, and they were on the lookout for strays.

"No," said Fritz. He rolled up his right sleeve. "See, nothing." The full head of hair and the healthy look of him were convincing enough, and the sergeant seemed satisfied. He put Fritz in the custody of a functionary prisoner who took him to the bathhouse.

There he met the Viennese prisoner again. This time he introduced himself properly; his name was Josef Kohl, though everyone called him Pepi. He was clearly a man of some importance; Fritz learned later that he was the leader of Mauthausen's resistance. Feeling instantly at ease with him, Fritz finally admitted the truth. Some of it, anyway: the fact that he'd been in Buchenwald and Auschwitz, and the story of his escape from the transport, right up to his arrest. That was all. He'd been a political prisoner, he claimed. Any hope of surviving here would depend on hiding his Jewishness.

For the third time, Fritz went through the ritual of being a new prisoner: the shower; clothes and belongings confiscated. When the clippers ran over his scalp and his new-grown hair fell in clumps, he knew he was back in the nightmare for good.

"Paying the price for keeping your address secret," said the Gestapo clerk as his details were taken down. Fritz looked enquiringly at him. "Only reason you're here," said the clerk, nodding at a note from the Wehrmacht officer on his desk. Seeing Fritz's expression, he added, "Too late for that now, my lad."

Did they still think he was a spy? Fritz was in a hideous dilemma. If he confessed the truth, there would be no hope of finding a way out of this situation. The sight of that quarry confirmed all he had ever heard about Mauthausen's evil reputation. But if he kept silent, he would be tortured and probably shot.

He decided it would be safest to confess, sticking to the same half-truth he'd given Pepi Kohl. Admitting he'd gone astray from the Auschwitz transport, he rolled up his left sleeve and revealed the tattoo. "Grounds for imprisonment?" the clerk asked.

"Protective custody," Fritz answered. "German Aryan, political."

The clerk didn't bat an eye. Fritz was entered on the records and assigned his third prisoner number: 130039.[2] No enquiries could be made about him, even if the Gestapo had been interested in doing so. Auschwitz no longer existed; it had finally fallen to the Red Army on January 27—the same day Fritz boarded the soldiers' train at Blindenmarkt. The only souls found in Monowitz had been the few hundred half-dead specters in the hospital, many of whom didn't survive long after liberation.[3]

Fritz gave the name of his cousin Lintschi—who was officially Aryan—as his next of kin, and his real Vienna address. As far as he knew from Fredl Wocher, there was nobody living there to be endangered by association with him. As for his trade, he calculated his answer. He'd acquired a variety of skills in the camps, but which ones should he admit to? It didn't look like there'd be much call for builders here, and he guessed that any surplus workers would end up in the quarry. So he told them he was a heating engineer.[4] It was half true— he'd helped build and fit out a few heating plants, and had learned from his papa how easy it could be to bluff one's way in a trade.

Although his escape bid had failed, it had at least given him a respite in which he'd built up his health and strength. He knew well what an advantage this would give him in surviving. What he didn't know was just how crucial that would be. Even after spending all his adult life in a hell on earth, the worst was yet to come.

בֵּן

Fritz was assigned to a block unsettlingly close to the camp bunker, which had a gas chamber and crematorium attached. In the next section of the camp, separated by a wall, hundreds of Soviet prisoners of war were kept in appalling conditions, starved and put to murderous hard labor. There had been a major escape two weeks earlier, the Russians using wet blankets to short the electric fence. Many were machine-gunned, but four hundred got out. For days afterward the local people had heard gunshots from the woods as the Russians were hunted down and executed.[5]

The camp was overcrowded, with blocks intended for three hundred prisoners holding many times that number. Like all concentration camps on Reich soil, Mauthausen was overflowing with evacuees from Auschwitz.

Fritz was looking forward to his reunion with his papa and friends, who must be somewhere among the multitude. But as he asked around, he couldn't find anyone who knew where they were, or who recognized their names. As far as he could gather, although there had been transports from Auschwitz, nobody knew of any that had arrived on or about January 26.

Eventually he had to conclude that his papa simply wasn't here, and never had been. But if that was the case, where on earth was he? Fritz had heard stories of atrocities in Poland and the Ostland—whole transports of Jews murdered in the forests. Was that what had happened to the Auschwitz transport? Was that the fate Fritz had escaped?

אַבָּא

Gustav sat with his back against the wagon wall. Fritz was gone, launched over the side into the freezing night. Pray God he would find his way to home and safety. Gustav was desperately weak and tired. He'd had no food for days, and only a mouthful of snow for moisture. "One man will kill another for a little scrap of bread," he wrote. "We are veritable artists of hunger . . . we fish for snow with a mug tied to a string dangled out of the wagon."

Later that night the train with its freight of dying men and corpses pulled in at the Mauthausen ramp. An SS cordon surrounded it. Hours ticked by; dawn came, and then the morning wore away. Inside the wagons, the men who still had their wits wondered what was happening. There appeared to be some kind of dispute going on.

A team of prisoners from the camp came along the train and handed out bread and canned food. There was little of it—half a loaf and one can between five men. It was devoured with terrifying savagery.

Eventually, with night drawing in again, the train began to groan and move, heading back the way it had come. Mauthausen's commandant, with his camp full to bursting, had refused to receive it.[6] It crossed the Danube again and turned west, in the direction of the German border. In a matter of hours they would be in Bavaria, and if the train carried on in a straight line it would bring them to Munich. That could mean only one thing: Dachau.

Gustav became aware of voices raised in urgent debate. A dozen of his comrades—including several of the old Buchenwalders—had been inspired by Fritz's example, and were talking of escape. They appealed to Gustav and to Paul Schmidt, who had been Fritz's kapo in the Buna Werke and had helped conceal him after his faked death. But Gustav could no more face it now than when Fritz had tried to persuade him, and Schmidt also declined to go. As the train trundled out of Linz, twelve of them climbed the sidewall and leapt over.

Despite the scale of the exodus there were no shots. The SS seemed oblivious; if more prisoners had had the strength, the train might have reached its destination empty except for the corpses.

Passing into Bavaria, they veered due north. Not Dachau, then. Day followed night—and another, and another, and still Gustav clung on to life. On the fifth day since leaving Mauthausen, they were in the German province of Thuringia, not far from Weimar. The train kept steaming north, and on Sunday, February 4—two weeks to the day since leaving Gleiwitz—it pulled into the freight yard at Nordhausen, an industrial town on the southern fringe of the Harz Mountains.[7]

It was met by SS guards and a Sonderkommando from the nearby Mittelbau-Dora concentration camp. Gustav climbed over the sidewall with extreme difficulty. Once the living had helped each other down, the dead were lifted out. By the end of the disembarkation, 766 corpses lay stacked on the loading ramp.

Gustav had seen some terrible things, but this was among the worst. "Starved and murdered," he wrote in his diary, "some frozen to death, and the whole thing not to be described." Many of the survivors were hardly in better condition than the dead—around six hundred of them died in the two days following their arrival, out of just over three thousand who had survived the journey.[8]

Tucked in a fold in a wooded ridge north of the town, Mittelbau-Dora concentration camp was about the size of Buchenwald. It was overcrowded, with over nineteen thousand prisoners crammed into its barrack blocks.

The new arrivals went through the registration procedure, Gustav receiving prisoner number 106498.[9] Assigned to blocks, they were fed at last—"the first warm meal since the start of our fourteen-day odyssey," Gustav noted. Each man got half a loaf, a portion of margarine, and a chunk of sausage, "on which we fell like hungry wolves."

Gustav remained in the camp for only two days before being selected for transfer to one of the smaller satellite camps. There were no transports, so they had to march the whole way, skirting the hill on which the main camp was built, and following the valley northwest to the village of Ellrich—a walk of eight and a half miles.

Ellrich concentration camp was by some margin the worst Gustav had yet experienced. It wasn't large, but it contained about eight thousand prisoners in wretchedly unsanitary conditions. Despite intakes from elsewhere, the population was constantly falling due to the death toll from starvation and disease. There were no washing or laundry facilities, and lice were endemic; a failed delousing program in the autumn had destroyed hundreds of prisoners' uniforms, which had never been replaced. When Gustav and his comrades arrived, they were confronted by the sight of filthy inmates, many of them in rags, some of them naked except for underwear. The "unclothed" were excused from work and restricted to half rations; as a result they were rapidly starving to death.[10]

Gustav's group were given two days' rest, then put to work.

Weakened by age and the wear and tear of five and a half years in the camps, plus the torment of the journey from Auschwitz, Gustav was shattered by the sheer unmitigated hell of Ellrich. It got to him in a way that nothing ever had before.

Every day, reveille came at 3:00 a.m., which in the depth of winter felt like the middle of the night.[11] The reason for this unholy start quickly became apparent. After a typically drawn-out roll call, the work details marched to the railway that ran by the camp and boarded a train that traveled to the village of Woffleben, where the main work site was located in a series of tunnels bored into the roots of the hills.[12]

Germany, under constant bombardment from the air, had moved much of its armaments production underground. In the Woffleben tunnels—carved out at appalling human cost by prisoner labor—they manufactured V-2 missiles, the most advanced and most terrifying of Hitler's secret weapons. The site resembled a quarry, with stepped cliff faces cut into the hillside; at the base, great openings like the entrances to aircraft hangars had been excavated. The whole outer area of the tunnel complex was covered with scaffolding elaborately draped in camouflage. The work that went on inside, in the depths of the earth, was top secret and, for the forced laborers, an unimaginable hell.

Gustav was drafted into a labor detail delving new tunnels, just to the west of the main complex. He was put with a group consisting mostly of Russian prisoners of war, doing the backbreaking work of laying railway tracks into the tunnels. The kapos and engineers were truly demonic slave drivers, lashing out with canes at anyone and everyone who caught their eye. Gustav had known nothing like it since the Buchenwald quarry. This was worse, because he had to suffer it without friends and on rations that wouldn't sustain a bed-bound invalid: two bowls of thin soup a day, with a piece of bread. For two whole weeks the bread issue stopped and they had to make do with just the watery soup, on which to endure a shift lasting from dawn until seven thirty in the evening. He lived in filth, and within weeks was as wasted and riddled with lice as the rest.

Ellrich was run by SS-Sergeant Otto Brinkmann, a little weasel of a man who was both a sadist and unfit for command. The commandant of Mittelbau-Dora had treated Ellrich like a dustbin into which he dumped his unwanted SS personnel and those prisoners who were least likely to survive. At evening roll call, when the prisoners were exhausted to the point of collapse, Brinkmann forced them to do exercises, lying down on the sharp stones of the unmade parade ground.

By Gustav's reckoning, fifty to sixty people a day were dying of starvation and abuse—"the perfect bone mill." But there was a grit in him that even now would not submit. "One can scarcely drag oneself along," he wrote, "but I have made a pact with myself that I will survive to the end. I take Gandhi, the Indian freedom fighter, as my model. He is so thin and yet lives. And every day I say a prayer to myself: *Gustl, do not despair. Grit your teeth—the SS murderers must not beat you.*"

He thought of the line he'd put in his poem "Quarry Kaleidoscope" five years earlier:

> Smack!—down on all fours he lies,
> But still the dog just will not die!

Recalling that image of resistance now, he wrote: "I think to myself, the dogs will make it to the end." His faith in that outcome was a rock, as firm as his belief that his boy was safe. Fritz, he was certain, must have reached Vienna by now.

בן

Fritz looked despondently at his food: a hunk of bread not much bigger than his hand and a small bowl of thin turnip stew. That, plus a mug of acorn coffee, was meant to sustain him through the whole day's labor. Sometimes he got extra stew, but it wouldn't hold his soul to his body for long. Just over a month had passed since his arrival, but his wrists were already visibly thinner and he could feel the sharpening of the bones in his face. He had never felt so abandoned, so devoid of friendship and support. Those bonds that had sustained him through Buchenwald and Auschwitz were no longer there; he had cut them away when he jumped from the train.

He was now in a subcamp at the village of Gusen, two and a half miles from Mauthausen. The events that had brought him here were, in their way, even stranger than those that had brought him to Mauthausen in the first place. With Germany fighting for its very existence and desperately short of men, the camp commandant, SS-Colonel Franz Ziereis, had announced that German and Austrian prisoners who were of Aryan blood could earn their freedom by volunteering for the SS. They would form special units, provided with uniforms and weapons, and would fight alongside the regular SS for the survival of the Fatherland.[13]

At a meeting of the Mauthausen resistance, Pepi Kohl and the other leaders had agreed that some of their people should volunteer. They guessed that the SS would attempt to use these units as cannon fodder or turn them against their fellow prisoners.[14] By infiltrating resisters into their ranks, the SS's own scheme could be turned against them; at the crucial moment, the volunteers would turn their weapons on the regular SS.

Among the 120 "volunteers" Pepi chose was Fritz. He was officially Aryan, healthy, and had the air of a fighter. Fritz was deeply reluctant; the very thought of putting on an SS uniform for any purpose sickened him. But Pepi was insistent, and wasn't the sort of man to be easily denied. And so it was that Fritz Kleinmann, Viennese Jew, had gone along to the commandant's office and signed up for the SS Death's Head special unit.[15]

The volunteers were taken to a nearby SS training school, where they began a hasty program of indoctrination and instruction. While the others managed to focus on their goal and reconcile it with what they were doing, Fritz found that he couldn't. The whole thing felt so profoundly wrong that he decided he had to leave. Resigning was impossible, so he began to misbehave, hoping to be kicked out. It was a dangerous tactic—potentially a path to a bullet in the back of the head. Eventually, after various punishments for minor infractions, he was dismissed from the unit. He became a prisoner again and was sent back to the camp, his SS career over before it had properly begun.

He was transferred to the subcamp at Gusen, one of a batch of 284 skilled workers, all perfect strangers to whom he felt little attachment. They were a cosmopolitan selection—Jews and political prisoners from all over the Reich: Polish, French, Austrian, Greek, Russian, Dutch; electricians, fitters, plumbers, painters, metalworkers, and general mechanics, plus one solitary Ukrainian aircraft mechanic.[16]

Gusen II held around ten thousand prisoners, many of them technical workers employed in secret aircraft factories in tunnels bored under the hills.[17] Fritz was assigned to labor battalion Ba III, a code name for a subunit working in the B8 "Bergkristall" aircraft plant in the tunnels by St. Georgen, where Messerschmitt built fuselages for its ultra-advanced Me 262 jet fighter.[18]

Fritz felt utterly isolated and friendless. Despondency took hold of him, as it had briefly in Monowitz. He scarcely noticed the passage of days through March and April; they remained in his memory only as a hellish blur.

The prisoners in the tunnels wasted away through starvation, while the SS and the green-triangle kapos murdered them at will. During March alone, nearly three thousand were declared unfit for work and despatched to Mauthausen, where most of them died. When a truckload of food was delivered to the camp by the International Committee of the Red Cross, the SS plundered it, taking the best for themselves, then pierced the remaining cans of food and condensed milk; laughing, they threw the leaking cans among the prisoners. Yet despite the death rate, the population grew rapidly as more and more death marches from evacuated camps across Austria were brought in.[19] They died in their thousands, and their unburied corpses piled up in the camps.

Physically as well as mentally Fritz altered. The conditions in Mauthausen-Gusen eroded him in two months from the lean young man he'd been when he left the Wehrmacht barracks in St. Pölten, whittling the flesh from his bones, until by late April he resembled the spectral, skeletal *Muselmänner.* The world in which he lived was the worst he had ever seen. Nothing but time—and a brief period at that—separated him from becoming another bony corpse on the heaps.

And yet, depressed as he was, he still didn't give up entirely as the *Muselmänner* did. There was an end in sight, if only he could cling on long enough to see it. Sounds of war were approaching—the familiar thumping of artillery in the far distance. The Americans were on their way.

The SS had planned for this. The Nazis had no intention of letting their top-secret jet-fighter production facility be captured—or their thousands of prisoners. On April 14, Heinrich Himmler sent a telegram to all concentration camp commandants: "No prisoner may fall alive into the hands of the enemy."[20] In Himmler's mind, that meant evacuation, and his telegram said as much. But in the mind of Mauthausen commandant Franz Ziereis, it was understood to mean a total liquidation. He laid his plans accordingly.

On the morning of April 28, all the prisoners in Gusen were held back from going to work. At ten forty-five the air-raid sirens sounded. Instantly, the SS and kapos began urgently herding the tens of thousands of prisoners toward the Kellerbau tunnels, the second set of underground works at Gusen.[21] They filed in through one of the three entrances—a huge maw as wide and high as a railway tunnel.

Inside, the granite and concrete walls were danker and colder than the sandstone of the Bergkristall tunnels. Due to the expense of excavating such hard stone and their vulnerability to flooding, the Kellerbau tunnels had never been completed,[22] but they were convenient as an air-raid shelter for the camps.

Fritz stood in the damp chill and waited, listening for the sounds of bombers and the thump of explosions. The minutes passed, but no sounds came.

He had never been in the Kellerbau tunnels, but some of those who had might have noticed as they filed in that two of the three entrances had been bricked up, leaving only this one open. Even the sharpest-eyed were unaware that, after they had entered, SS machine gunners had set up positions outside. The prisoners were also ignorant of the fact that during the previous few days, this last entrance had been mined with explosives on orders from Ziereis. The operation was code-named *Feuerzeug*—Lighter.

The task had been undertaken by the civilian manager in charge of tunnel construction, Paul Wolfram; he and his colleagues were told that their own and their families' lives would be in jeopardy if they botched the job or revealed the secret.[23] Wolfram had laced the entrance with all the explosives he had at hand. It was insufficient, so he added a couple of dozen aerial bombs and two truckloads of marine mines. During the night before the air-raid alert, the explosives had been wired up.

Now, with all the prisoners inside and the machine gunners ready to prevent escape, the tunnel entrance was ready to be blown. The inmates, sealed inside, would suffocate to death.

20. The End of Days

By the close of March, when he'd been at Ellrich about a month and a half, things had improved a little for Gustav; just enough to nourish his will and keep his body and soul together.

He'd been taken off track-laying and was working in the tunnels as a carpenter. His kapo was a decent man named Erich, who had secret sources of food for himself and donated his soup ration to Gustav[1]—just enough to slow but not reverse the process of starvation. Meanwhile he grew more filthy and lice-infested with every passing day.

Gustav lived his days underground. It was like the fourth circle of the pit of Dante's Hell: most of the laboring slaves were on the brink of death, the stronger preying on the weak, robbing them of their paltry rations. The only plentiful thing was corpses, and there had been occurrences of cannibalism. Over a thousand prisoners had died in March, and a further sixteen hundred walking skeletons had been sent to an army barracks in Nordhausen that served as a dump for the spent and useless.[2]

By April, American forces were only days away, and the SS began pulling the plug. Work was halted and preparations began for evacuation. That same night, the RAF firebombed Nordhausen, hitting the barracks and killing hundreds of sick prisoners. They visited again the next night, razing the town and adding more prisoners to the death toll.[3]

The evacuation of Ellrich took two days. Gustav and all the

other prisoners who were fit to move were loaded into cattle wagons. As the final train prepared to leave on April 5, the last SS man to depart the camp shot the dozen or so remaining sick prisoners. When the US 104th Infantry Division reached Ellrich a week later, they found not a living soul.[4]

אבא

Gustav thought back on the journey from Auschwitz. The weather was milder now, he had room to sit, and they even got a little food. Not nearly as much as they should have received, however; supply wagons stocked with food had been coupled to the rear of the train when it left Ellrich, but at some point they had been disconnected. A little relief came when the train stopped at a town where there was a bread factory.[5] A British prisoner of war gave Gustav four and a half pounds of pumpernickel and other breads—enough to keep him and his immediate comrades going for three days.

The train had come far into the north of Germany, past Hanover, and on April 9 it reached its final destination: the small town of Bergen, the unloading place for Bergen-Belsen concentration camp.

With the ring of enemies closing in ever more tightly, Himmler was determined to hold on to his surviving prisoners. They were intended to serve one final purpose—as hostages. Bergen-Belsen was one of the last concentration camps remaining on German-held soil. By the time Gustav arrived, the camp, designed for only a few thousand, had swollen beyond all sense or reason, and despite thousands of deaths every month from starvation and disease—7,000 in February, 18,000 in March, 9,000 in the first days of April—the living population had climbed to over 60,000 souls, existing among piles of unburied corpses in an atmosphere rife with typhus. In

Himmler's peculiar mind, he was saving them, trying to win favor with the Allies by showing how merciful he was to the Jews, rather than the architect of their mass murder.[6]

Into this boiling mass of humanity, Gustav and the other survivors of Ellrich were to be driven.

Many had not survived the journey, and there was the usual cargo of corpses to be unloaded from the train. As the survivors marched from the station toward the camp, an astonishing thing happened that was both terrible and wonderful. The column of ghosts met another marching in the same direction; they were Hungarian Jews—men, women, children—all starving and wretched. Many of the Ellrich survivors were Hungarian also, and to Gustav's amazement, first one person then another and another from one column recognized relatives in the other. They broke ranks and ran to them, calling their names. Friends, mothers, sisters, fathers, children, long separated and thinking their dear ones dead, found them again on the road to Belsen. It was both joyous and heartrending, and Gustav could not find the words to describe what he saw—"one can only imagine such a reunion." What would he not have given to be so reunited with Tini and Herta and Fritz. But not here, not in this place.

There were no anchors left, no touchstones, no certainties; even the camp system had broken down. With Belsen full to bursting, the fifteen thousand who arrived from the Mittelbau camps were turned away. Their SS escorts found accommodation for them at a nearby Wehrmacht panzer training school between Belsen and Hohne. Its barracks were pressed into service as an overflow concentration camp, designated Belsen Camp 2, under the command of SS-Captain Franz Hössler, who had accompanied the transports.[7] A thuggish-looking individual with a jutting chin and sunken mouth, before Mittelbau, Hössler had commanded one of the women's sections in Auschwitz-Birkenau, participating in selections and gassings and countless acts of individual murder. It had been Hössler who had selected the women "volunteers" for the Monowitz brothel.[8]

Physically the panzer training school was a pleasant change for the prisoners: clean, airy white buildings set around tarmac squares dispersed among woodland. The Wehrmacht staff—now consisting of a Hungarian regiment—helped the SS guards manage the prisoners.

The rations improved in quality, but the quantities were pathetically inadequate. Gustav and his comrades were reduced to foraging potato and turnip peelings from the rubbish bins outside the barrack kitchens—"anything to relieve the hunger," he noted in his diary.

In all Gustav's time in the camps, he had never been surrounded by so much tight-pressed humanity, or seen helpless starvation on such a colossal scale. After all he had endured, here in Belsen the faith in himself that had kept him going started to waver. What made *him* special? Why should *he* make it to the end when so many millions had not or would not?

In their own way, the Hungarian troops were as brutal as the SS. Most of the officers were well groomed, with pomaded hair, and had instilled in their mostly illiterate men an anti-Semitic fascist ideology that was on par with anything the SS could provide. They were callous and apt to shoot inmates for entertainment. Their main duty was to protect the kitchens, and they would stand in the square between the barracks taking shots at the prisoners foraging for scraps, killing dozens of them.[9] Some retained a mystic devotion to the Nazi cause. One told a Jewish woman that he regretted the work of exterminating her people remained incomplete, telling her that Hitler would surely return, "and again we shall fight side by side."[10]

On his first night in Belsen 2, Gustav stood vigil in the

upper story of his building. In the south he saw the dark sky glowing orange. It looked to him as if a town—possibly Celle, twelve or so miles away—was in flames. Even as Gustav watched, it flashed and erupted with explosions. This wasn't aerial bombing—it was a battlefront.[11]

His sinking heart began to rise. "I think to myself, now the liberators must be here soon—and I have faith again. I think to myself still, the Lord God does not forsake us."

Two days later, on April 12, local Wehrmacht commanders made contact with British forces and negotiated for the peaceful surrender of Bergen-Belsen. In order to contain the epidemic of typhus, a zone of several miles around the camp would become neutral territory.

In the barracks, Gustav noticed that most of the Hungarian soldiers had begun wearing white armbands as a token of neutrality. Even some of the SS were doing the same—including the camp leader, SS-Corporal Sommer, whom Gustav had known in Auschwitz as "one of the bloodhounds." It seemed that the prisoners would be handed over to the British without bloodshed. "It is high time," Gustav wrote, because the SS "wanted to make of us a St. Bartholomew's Night massacre under English illumination, but the Hungarian colonel didn't want any part of it, and so they have left us alone."

On April 14, Gustav saw the first British tanks in the distance. The word spread through the barracks, bringing joy at the news, and the celebrations went on all night.

חברים

Captain Derrick Sington struggled to make himself heard over the convoy of tanks clanking and roaring through the town of Winsen. Following a race to catch up with the armored vehicles of the 23rd Hussars, Sington had found the regiment's intelligence officer, and was trying to inform him of his special mission over the din of military traffic.

Derrick Sington was commander of No. 14 Amplifying Unit of the Army Intelligence Corps. Equipped with light trucks mounted with loudspeakers, their role was to disseminate information and propaganda. His orders were to accompany the advance column of the 63rd Anti-Tank Regiment, which would be establishing the neutral zone surrounding the Bergen-Belsen camp. The prisoners—or "internees" as the British were officially calling them—must not be allowed to leave the zone, due to disease. Captain Sington's urgent mission was to locate the camp and make the requisite announcements to the inmates. As a German speaker, he would also act as interpreter for Lieutenant Colonel Taylor of the 63rd, who would be in overall command of the zone.[12]

Yelling at the top of his voice over the clattering of tracks and roaring engines, Sington explained all this to the Hussars officer, who leaned out of the turret of his tank with his hand cupped over his ear. He nodded and told Sington to fall into line. Sington jumped back into his seat, gestured to his driver, and they pulled into the road, joining the flow of armor.

Beyond Winsen, the column passed through open countryside, which gave way to thick woodlands of firs, whose powerful scent mingled with the exhaust fumes and the stench of burning. The infantry were torching the undergrowth on either side with flamethrowers; they weren't taking any chances with concealed German anti-tank weapons or snipers.

Not far up the road, Sington saw the first warning notices—"DANGER TYPHUS"—marking the perimeter of the neutral zone. Two German NCOs passed him a note written in bad English inviting him to meet the Wehrmacht commandant at Bergen-Belsen.

As the road swung eastward Sington spotted the camp—an

enclosure of high barbed-wire fences and watchtowers among forest, flanking the side of the road. He was met at the gate by a small group of very smartly dressed enemy officers: one in the field gray of the Wehrmacht, a highly decorated Hungarian captain in khaki, and a bulky, fleshy-faced SS officer with a simian jaw and a scar on his cheek, who proved to be SS-Captain Josef Kramer, the former commandant.

While they waited for the arrival of Colonel Taylor, Sington fell into polite conversation with Kramer. He asked him how many prisoners were in the camp; Kramer answered forty thousand, and an additional fifteen thousand in Camp 2 up the road. And what kind of prisoners were they? "Habitual criminals and homosexuals," said Kramer, looking furtively at the Englishman. Sington said nothing, but later noted that he had "reason to believe it was an incomplete statement."[13]

Their conversation was mercifully cut short by the arrival of Colonel Taylor's jeep. He ordered Sington to go in and make his announcement, then roared on up the road toward Bergen. At Sington's invitation Kramer climbed up on the running board of the loudspeaker truck, and they drove in through the gates.

Sington had tried many times to imagine what the inside of a concentration camp would be like, but it was unlike anything he had pictured. There was a road through the center, with separate compounds on either side, each filled with wooden barrack blocks. The place was suffused with "a smell of ordure" which reminded Sington of "the smell of a monkey-house" in a zoo; "sad blue smoke floated like a ground mist between the low buildings." The excited inmates "crowded to the barbed wire fences . . . with their shaven heads and their obscenely striped penitentiary suits, which were so dehumanising." Sington had witnessed gratitude from many different liberated peoples since Normandy, but the cheers from these skeletal, wasted ghosts, "in their terrible motley, who had once been Polish officers, land-workers in the Ukraine, Budapest doctors, and students in France, impelled a stronger emotion, and I had to fight back my tears."[14]

He stopped his truck at intervals, the loudspeakers blaring out the announcements that the camp zone was in quarantine under British administration; the SS had surrendered control and would now withdraw; the Hungarian regiment would remain, but under direct command of the British army; prisoners must not leave the area due to the risk of spreading typhus; food and medical supplies were being rushed to the camp with all haste.

The joyful inmates spilled out of the compounds and surrounded the truck. Kramer was alarmed, and a Hungarian soldier began firing directly over the heads of the prisoners. Sington jumped out of his truck. "Stop shooting!" he ordered, pulling his revolver, and the soldier lowered his rifle. But no sooner had he stopped than, to Sington's amazement, a band of men in prisoner uniforms armed with cudgels ran into the crowd, lashing and beating with appalling brutality.

Arriving back at the main gate, Sington said to Kramer, "You've made a fine hell here."[15] His brief tour had shown him only the throng of survivors, and it would be a day or two before he finally discovered the burial pits, the crematorium, and the grounds strewn and stacked with thousands of naked, emaciated corpses.

Pulling out of the gate, he turned his truck toward Camp 2, to repeat his round of announcements.

אבא

A day had passed since Gustav had seen the tanks in the distance. At long last the British column came rolling up the main

Bergen road, passing by the camp altogether. Little seemed to happen. Then the loudspeaker truck arrived in Camp 2. The prisoners gathered round to hear the British officer's announcement, which was drowned out by cheering.

The prisoners in Camp 2, although in a dreadful state, weren't nearly as wretched as those in the main camp. They had strength left and they had anger. As soon as Captain Sington's truck had departed, the lynchings began.

Hundreds of men, exalted in their fury and their strength in numbers, singled out the individuals who had tortured them. Gustav—the kindest, gentlest soul imaginable—watched dispassionately as SS guards and green-triangle block seniors were strung up or beaten to death. He saw at least two murderers from Auschwitz-Monowitz die, and felt no pity or remorse. The Hungarian troops made no move to intercede. That afternoon, when the killing was done, the surviving SS were made to remove the bodies, burying them the next day with their own hands.

The British gradually took control of the administration, creating records of all the surviving prisoners, ordering them by nationality; Bergen-Belsen was transformed into a displaced persons camp, and the inmates were prepared for repatriation. Gustav remained with the Hungarian Jews; he'd made many good friends among them, and they'd elected him room senior.

It was a liberation and yet not a liberation. The inmates were no longer under the heel of the SS; the British brought in food and medical supplies, and they ate well and began to recover their health (although in the main camp the inmates were in such a bad state that thousands died in the weeks following liberation). Yet they were still prisoners. The Hungarian soldiers were under orders from the British to prevent anyone from leaving, and took the orders seriously. As far as Camp 2 was concerned, the quarantine was preposterous—there was no typhus here, and no need to keep the prisoners incarcerated. Gustav began to chafe, longing to experience freedom again after all these years.

The liberation of Belsen made headlines around the world; there were newsreels and radio reports, and the papers were full of it. Across Europe and in Britain and America, the relatives of people in Nazi captivity sent desperate requests for information. Periodically, Captain Sington's loudspeaker truck toured Camp 2, broadcasting the names of people whose families had enquired.[16]

Gustav thought of Edith and Kurt. He hadn't seen his daughter since her departure for England in early 1939, and had heard no news of her since the start of the war. Kurt too had been cut off since December 1941. Gustav wrote a message to Edith detailing his whereabouts and block number, and entrusted it—along with the thousands of messages from other inmates—to the British administration.[17]

In the main camp, medical staff worked to save as many lives as possible. It was a place that blasted the minds of those who witnessed it. The corpses were heaped in thousands, and the half dead, half living moved around them as if they were just lumber, stepping over them, sitting down to eat their scraps leaning against stacks of corpses.[18] Deep pits were excavated, dozens of feet long. SS guards were made to carry the dead into the pits by hand, jeered at and cursed by survivors; a few SS men made a run for the forest, but they were shot down and their comrades had to drag their bodies back. Into the pits they went, along with their victims.[19] The task proved overwhelming in the end; there were just too many bodies, and bulldozers had to push the decomposing corpses into the pits. It took nearly two weeks before the last were buried.[20]

The survivors of Camp 1 were moved to the clean, solid buildings of Camp 2, which was redesignated as a hospital. As the insanitary, broken-down wooden barracks in the main camp emptied, they were torched with flamethrowers.

An English nurse on the medical staff felt shame and remorse that, having heard of the existence of such camps as early as 1934, she had never realized—and hadn't wanted to realize—that they could be like this. She and her colleagues were "stirred with a cold anger against those primarily responsible, the Germans, an anger which grew daily at Belsen."[21] Others were shocked by how abuse and degradation had reduced many survivors to an animallike state, fighting for food, eating from bowls that doubled as bedpans, with only a wipe with a rag between uses.[22]

The influx from the main camp raised a problem for Gustav and the Mittelbau survivors: it brought typhus into their vicinity. Buildings housing the infected were cordoned off, but their presence still increased the risk that it would spread throughout the barracks.

Gustav was growing desperate to leave this terrible, haunted place.

Ten days after the liberation, the first repatriation transports were allowed to leave, carrying a selection of French, Belgian, and Dutch survivors. Their way home lay through liberated countries. Those who came from Germany, Austria, and other places that were either war zones or still in German hands would have to wait. Gustav watched the transports go with longing, and as the days passed he lost patience. It didn't matter that it was irrational, that Austria wasn't yet free; he was sure he could find his way home, no matter what. He believed that Fritz would be in Vienna now, waiting for him. Gustav needed to get back to him somehow. At the very least he yearned to be free of confinement.

Choosing his moment, on the morning of Monday, April 30, he set out. Taking his few belongings and a little food, he walked out of his building and along the tarmac path, heading for the road.

A Hungarian soldier stepped in front of him, rifle raised.

"Where d'you think you're going?"

"Home," said Gustav. "I'm leaving." There was a look in the soldier's eyes that Gustav had seen in hundreds of SS guards— the look of an anti-Semite regarding a Jewish prisoner. Until two weeks ago this soldier had been fighting alongside Nazis. Gustav went to walk past him. The soldier swung his rifle butt and smacked him in the chest. Gustav staggered, wheezing.

"Try that again and I'll shoot you," said the soldier.

Gustav had seen enough to know he wouldn't hesitate. The bid for freedom was over; he was trapped.

Nursing his bruised ribs, he walked back to his block. Getting out of Belsen would be trickier than he'd anticipated. He talked it over with a fellow Viennese, a man named Josef Berger, who was also desperate to go home.

That afternoon, the two men left their building quietly and hung about, keeping an eye on the sentries. At last came the moment they were waiting for, when one guard shift was relieved by the next. While the soldiers were distracted, Gustav and Josef made a dash for it—not toward the road this time but in the direction of the woods fringing the northwestern edges of the camp.

They were between sentry posts when there was a shout in Hungarian and the crack of a rifle. The bullet snapped over their heads. Another shot zipped past, and they both threw themselves flat on the ground. Bullets thwacked into the turf around them. Gustav and Josef crawled along on their elbows. As soon as there was a pause in the shooting, they jumped to their feet and made a break for the woods, dodging, hurling themselves through the trees and out the other side. They ran on through the Russian section of the camp and into the forest on the far side.

בן

Minutes ticked by in the chilly, dripping Kellerbau cavern, but there were no sounds of planes, no thump of explosions: all around Fritz, there was just the echoing susurrus of tens of thousands of prisoners breathing and muttering to one another.

Time ticked on. Outside the tunnel, the SS machine gunners watched and waited for the imminent explosion.

Minutes turned slowly into hours. The prisoners, accustomed to standing at roll calls that dragged on just like this, thought little of it. A false alarm, presumably. At least they weren't having to work. Most of them would never learn the true reason why they'd been sent into the tunnels, and would never be aware of the complications that kept them standing there for so long. Events were going on over their heads, veiled in obscurity, that would never be fully revealed.

The explosive charges embedded around the entrance had failed to detonate. Paul Wolfram, the manager in charge, would later claim to have deliberately sabotaged the murder plot by having his men plant the bombs and mines without detonators. But that didn't explain the other explosives. Commandant Ziereis—who spent much of this period drunk—claimed that he had reservations about the whole business. But a story circulated among some of the survivors was that a Polish prisoner named Władvsłava Palonka, an electrician, had discovered the detonator wires and cut them.[23]

At 4:00 p.m. the all-clear sounded, and the prisoners who had walked in unknowingly to their deaths walked out again— still with no idea of the death sentence that had hung over them—and filed back to the camps. Had the plot succeeded, it would have killed over twenty thousand people—one of the largest single acts of mass murder in the history of Europe.[24]

The routine of roll calls and labor resumed, but then, on

Tuesday, May 1, the prisoners were not sent to work. Fritz sensed a mood among the SS that reminded him of Monowitz in the middle of January. This time the panic was deeper. The SS now had no Reich left into which they could retreat. There would be no evacuation from Mauthausen.

Two days later, all the guards disappeared from the camp. The fanatical Nazis among them intended to fight a last-ditch defense in the mountains, while the rest shed their uniforms and went into hiding among the civilian populations in the cities. Command of Mauthausen-Gusen was officially handed over to the Vienna civil police force, while camp administration fell to the Luftwaffe. A detachment of the Vienna fire brigade who had come here as political prisoners in 1944 were drafted to help.[25]

To the south, armies made up of Americans, British, Poles, Indians, New Zealanders, and one Jewish Brigade were pushing into the mountainous borderland between Italy and Austria.

In the east, the Red Army had crossed the Austrian border and by April 6 had surrounded Vienna. The German forces remaining there were hopelessly insufficient to defend it, and the siege was short-lived. By April 7, Soviet troops were in the southern part of the inner city, and three days later the Germans evacuated Leopoldstadt. The Danube bridges were captured, and on April 13, the last SS armored unit abandoned the city.[26] Vienna was liberated, seven years almost to the day since Hitler had held his plebiscite and set the seal on the Anschluss. Now he was trapped in his Berlin bunker and his grand Reich was reduced to a tiny, bleeding stump.

The third Allied spearhead came from the northwest. American forces crossed the Bavarian stretch of the Danube on April 27. Patton sent his XII Corps into Austria north of the river. They faced heavy fighting from fanatical German forces who had taken to hanging deserters from roadside trees.[27] As the American advance pushed down the Danube valley, the very tip of the spear consisted of a patrol from the 41st Cavalry Reconnaissance Squadron and a platoon from the 55th Armored Infantry Battalion. Probing east of Linz they came to the villages of St. Georgen and Gusen, where they first laid eyes on the camps.

For sheer horror, Mauthausen and Gusen rivaled Bergen-Belsen. Both had been sinks into which the concentration camps had drained. The death rate at Mauthausen had spiraled to over nine thousand a month. The walking cadavers who greeted the American liberators were found to be living among tens of thousands of their unburied, half-buried, or half-burned dead. The stench of it was what stuck in the minds of the GIs. "The smell and the stink of the dead and the dying, the smell and the stink of the starving," recalled one officer. "Yes, it is the smell, the odor of the death camp that makes it burn in the nostrils and memory. I will always smell Mauthausen."[28]

Olive-green tanks marked with the white American star, scarred and weathered, rolled into the camp compounds. In Gusen I, a sergeant stood on top of his Sherman and yelled in English to the crowd of emaciated prisoners, "Brothers, you are free!"[29] Bursts of various national anthems exploded from the crowd, and the Volkssturm officer in command of the German guards presented his sword to the sergeant.

Fritz, in the neighboring Gusen II, watched the Americans arrive with relief and satisfaction, but without any overwhelming joy. He was too weak and demoralized to celebrate. Having been only passably healthy when he arrived, he had endured three months in this hell, where life expectancy even for the fittest new arrivals was only four. He was scarcely alive, little more than bones wrapped in skin, covered with bruises and sores. He had no real comrades in Mauthausen-Gusen, only fellow sufferers. "I was utterly demolished there," he wrote later.[30] He was too weak and sick to go home—even assuming there was a home to go to. Most of all he missed his papa, but hadn't the faintest idea what had happened to him.

אבא

After a half mile or so, Gustav and Josef stopped to catch their breath. They listened, but there were no sounds of pursuit, just birdsong and the muffled silence of the wood. They sank down to rest. Gustav looked about him, gazed up at the sky, and inhaled the fir-scented air. The very smell of it gave him joy; it was the scent of liberty. "Finally free!" he wrote in his diary. "The air around us is indescribable." For the first time in years, the atmosphere was untainted by the odors of death and labor and unwashed human hordes.

They weren't safe yet. The front lines lay east, so for the time being Gustav and Josef turned their backs on their homeland and pressed on northwest through the forest.

All afternoon and into the evening they walked, passing several tiny hamlets scattered among the woods—German places, where they didn't dare ask for help. Eventually, after about twelve miles, they emerged from the forest into the small village of Osterheide. On the outskirts stood a large prisoner-of-war camp—Stalag XI—which had been liberated by the British the day after Belsen.[31] It had been evacuated several days ago, but there was still a population of Russian POWs, who gave the itinerant Viennese bed and board for the night.

The next morning, Gustav and Josef walked on into Bad Fallingbostel, a pleasant spa town, choked with refugees and troops. The two men presented themselves to the British authorities, but were told that nothing could be done for them right away—they ought to be in one of the displaced persons camps. They fared better at the German mayor's office, where they were assigned accommodation in a hotel and allotted a food ration.

Gustav found himself a week's employment as a saddler with a local upholsterer named Brokman. The wages were decent, and for the first time in seven years he was treated as a citizen. He began to recover from his ordeal. In his room at the hotel, he brought out the little green notebook that had accompanied him since the early days. On the first page was the entry: "Arrived in Buchenwald on the 2nd October 1939 after a two-day train journey. From Weimar railway station we ran to the camp . . ."

So began the record of his captivity. Now he started recording his liberty.

"At last one is a free man, and can do as one pleases," he wrote. "Only one thing nags at me, and that is the uncertainty about my family at home."

It would continue to prey on his mind, so long as the remnants of the Nazi regime remained, still fighting, in the territory between himself and his homeland.

21. The Long Way Home

משפחה

Edith stood at the front window, watching the postman wheel his bicycle up the hill. The grandly named Spring Mansions—a genteel three-story Victorian house converted to apartments—stood at the corner of Gondar Gardens in Cricklewood, from which half of London could be seen laid out: the railway, and beyond, the Kilburn High Road, cutting a straight line all the way to Westminster.

Little Peter stood beside her, looking out at the view. He had only just returned to his parents after being evacuated to a farm near Gloucester. During his absence his parents and his baby sister, Joan, had left Leeds and moved to this little apartment in London. Peter was almost a stranger to his mother—completely British by birth and accent. Edith and Richard, conscious of the hostility to all things German in this country, never spoke anything but English in the house.

The postman propped his bicycle on the hedge and dropped a bundle of letters through the letterbox. Edith went downstairs and collected them from the mat; flicking through the envelopes for the other tenants, she came across one addressed to "Fr. Edith Kleinmann." There were several crossed-out addresses, starting with Mrs. Brostoff's in Leeds. She tore open the envelope.

Peter heard his mother running up the stairs, calling breathlessly for his father. Peter couldn't make out what the excitement was about; she just kept repeating that her father was alive. *Alive.*

It was almost beyond belief. All this time, she'd had no notion of what had become of her family; she knew from Kurt that their father and Fritz had gone to Buchenwald, but that was all. Everyone had seen the dreadful newsreels from Belsen and heard the BBC broadcasts—and to think that her papa had been there and lived!

Edith wrote immediately to Kurt. On their behalf, Judge Sam Barnet used every political connection he had to try to open a line of communication with their father.[1] Weeks went by, and no further word came from Gustav. It was as if, having revealed his presence, he'd suddenly vanished.

בן

Following liberation, the US Army brought medical aid to the survivors of Mauthausen and Gusen. Thousands were beyond saving, and died in those first days.

Fritz Kleinmann was among those whose grip on life still held, despite his desperate condition. When the medical assessments began, he was interviewed by an American officer, who revealed that he'd been born in Vienna himself, in Leopold-stadt. Pleased by this connection between them, the officer found Fritz a priority place on an emergency evacuation.

He was taken to Regensburg in southern Bavaria, a beautiful, ancient city where an American military hospital had been set up. His arrival coincided with the news that Germany had surrendered; Hitler and Himmler were dead, and the war in Europe was over.

The 107th Evacuation Hospital was housed in tents and buildings on the bank of the river Regen where it flowed into the Danube.[2] When Fritz was checked in, he was scarcely alive, his weight recorded as seventy-nine pounds. The hideous, mi-raculous, haphazard chain of events that had allowed him to evade death for five and a half years had nearly finished him off at the end.

Resting on his cot in the hospital tent, he knew that the ordeal that had begun on that day in March 1938, when the Luftwaffe dropped their snowstorm of leaflets all over Vienna, was finished.

Except it wasn't, quite. The journey that had started that day would not be complete until he returned to Vienna and discovered whether it was still his home—and, most important, whether his father had survived. And as for the nightmare—why, that would never end so long as life and memory lasted. The dead remained dead, the living were scarred, and their numbers and their histories would stand for all time as a memorial.

Leaving the future to take care of itself for the moment, Fritz focused on regaining his strength. The doctors gave him a diet of cookies, milk pudding, and a strength-building concoction whose ingredients he never knew. Within two weeks he had gained twenty-two pounds. He was still severely underweight, but he felt strong enough to travel, and could feel the pull of home. The hospital was packing up to move to a new location, and his request to be discharged was granted. He went to the city hall in Regensburg, where he was issued with civilian clothes and listed for transport back to Austria.

It was late May when Fritz passed through Linz and arrived at the demarcation line between the American and Soviet zones on the south bank of the Danube, opposite Gusen and Mauthausen. At St. Valentin he caught a train. Yet again he took the railway journey through Amstetten, Blindenmarkt, and St. Pölten. This time he met with no interference.

At last, on Monday, May 28, 1945, Fritz set foot in Vienna, five years, seven months, and twenty-eight days since leaving it on the transport bound for Buchenwald. His train came in at the Westbahnhof, the very station from which he had departed. Fritz later discovered that of the 1,035 Jewish men who had been on that transport, only twenty-six were still alive.

Vienna hadn't suffered as much in the recent fighting as Berlin had. The siege had been brief, with no wholesale destruction. Parts of the city were scarcely touched. However, it was Fritz's luck that the route he walked from the station to the city was one of the worst hit, giving him the distressing impression that Vienna had been all but destroyed.

It was late in the evening, and the darkness of a summer night was settling over the streets when he reached the Danube Canal. The buildings on the Leopoldstadt side were badly damaged by bombs, and the once handsome Salztor Bridge was just a jagged stump protruding from the bank. Fritz crossed elsewhere and eventually found himself in the Karmelitermarkt.

The stalls had been packed away, the cobblestones were bare, and it was as it had been on those evenings long ago when he and his friends had played here, kicking a rag ball around, watching out for the police, being warned off by the lamplighters for climbing the lampposts. He could recall the cream cakes, the pink *Mannerschnitte* wafers, the bread crusts and ends of sausages, the shopkeepers and stallholders, Jews and non-Jews driving their trade side by side, thriving without hate or hostility, their children playing in a single rushing, laughing society. Now half of what had made this place live was gone: ashes from the Auschwitz ovens floating down the Vistula; bones in the soil under the pine needles of Maly Trostinets, or scattered to the world—Palestine, England, the Americas, the Far East. Aside from a handful like Fritz, they would never return to the Karmelitermarkt.[3]

When he reached the old apartment building in Im Werd, he found the outer door locked. The Soviet authorities had imposed a curfew that began at 8:00 p.m. He hammered on the door, and it was opened by the familiar figure of Frau Ziegler, the building caretaker. She greeted him with amazement. Everyone had thought he and his father were dead.

She let him in, but wouldn't allow him up to the old apartment; there were new people living there now who'd been bombed out of their home. There were no Kleinmanns here anymore.

On his first night back in Vienna, Fritz slept on Frau Ziegler's floor. When he rose next morning and went out, he found that the news of his return had preceded him. "The Kleinmann boy is back," they said to one another in astonishment.

He didn't see Olga Steyskal or any of his papa's other friends that morning, but he did run into Josefa Hirschler, the caretaker of Olly's apartment building. She greeted him warmly, and invited him to take his first Viennese breakfast with her and her children, who were old friends of his. He was begrimed from his journey across Austria, so Josefa sent him out to the back courtyard to clean up. He found a bowl waiting for him filled with hot water.

As he splashed his face and scrubbed at his neck, he felt that a new life was beginning for him. But it was a new life alone, without any family. His little brother in America, his sister in England, his mother and Herta gone, almost certainly dead in the east . . . As for his papa, there didn't seem much hope; he'd seemed near to death when they parted. *You have to forget your father* . . . Were Robert Siewert's words going to come true at last, here at the end of the road? If by some miracle his papa had survived, where in the world was he?

אבא

Gustav had found himself a good life in Bad Fallingbostel; he had work, and was eating well. He'd made friends with a

German woman from Aachen who gave him extra food. He made rucksacks for some Serbian army officers who had been prisoners of war; they seemed well supplied, and gave him lots of cigarettes.[4]

"I feel much stronger," he wrote, but "Dear Lord, if only I were in Vienna with my son." He had never doubted that Fritz had made it home after jumping from the train.

Several other Viennese drifted into Fallingbostel and formed a tiny community. With the war finally over, Gustav and his new friends at last set out on the long trek home.

They went slowly, finding food and shelter wherever they could, passing through the mountainous forest country south of Hildesheim. Gustav appreciated the slowness of the journey, relishing the freedom and the beautiful scenery. In the town of Alfeld, he bumped into an old friend who'd been a political prisoner in Buchenwald and was now chief of police, no less. Hearing about the journey Gustav had before him, he gave him a bicycle.

The pace picked up, and on May 20, the traveling party reached the city of Halle in Saxony, where Gustav was reunited with many more old comrades from both Monowitz and Buchenwald. Among the latter was his good friend and Fritz's mentor Robert Siewert, who had survived right to the end, and had come back to his old hometown to begin rebuilding its Communist Party.

Halle proved to be something of a gathering place for concentration camp survivors, and Gustav decided to stay for a while. They received good care and plenty to eat, and there was an established Austrian committee. Robert Siewert gave a public talk on conditions in Buchenwald—beginning his lifelong task of helping to keep the memory alive.

After a month in Halle, the journey resumed. Cycling through Bavaria, Gustav exulted in the beauty of nature. "This area is glorious," he wrote during one of the frequent halts. "Nothing but mountains everywhere. I feel as if I have been born again."

In late June they reached Regensburg, and on July 2 Gustav rode his bicycle over the Danube bridge at Passau and entered Austria, welcomed home by the pealing of church bells striking noon.

The Austrian exiles reached Linz after dark, in pouring rain. It was too late to find accommodation, so they passed the night in an air-raid shelter. Provided with ration cards, they spent several days in the city.

Although he was on home soil and Vienna was only a train ride away, Gustav's footsteps slowed again. After traveling so far, he suddenly felt no great urgency to reach home. He was enjoying himself, and although he never confessed as much to his diary, there must have been an anxiety nagging at the back of his mind that distressing news might be waiting for him. Not only the full truth about what had happened to Tini and Herta, but what if his faith was mistaken, and Fritz wasn't there?

More than anything he was relishing his freedom. For the first time ever—not only since the camps but for the first time in his whole life—Gustav was completely at liberty, without responsibility or cares or fear, free to go as he pleased and take his time drinking in the sights and smelling the flowers.

One day, taking advantage of the beautiful weather, he took a day trip up into the mountains with one of his companions.[5] Acting on an impulse, they went to the village of Mauthausen, where yet another old camp comrade, Walter Petzold from Auschwitz, was now chief of police. They walked up the hill and had a look at the concentration camp, its formidable stone enclosure now deserted. Gustav was curious to see the place from which the Auschwitz train had been turned away. Had he known that Fritz had spent three months here, and that it had nearly killed him, he might have regarded it in a different light.

On July 11, Gustav crossed the "green border" for the first time, passing from the American into the Soviet zone. He found the Russians "very courteous to us concentration camp survivors." Through the rest of July and August, he lingered in central Austria, and it was only when the summer began to wane that he finally steered his bicycle toward the last homeward leg.

On a day in September Gustav Kleinmann entered Vienna. He saw the devastation, the massive concrete flak towers looming over the pretty parks, and he saw all the familiar landmarks. The Karmelitermarkt was still there, and the apartment buildings of Im Werd overlooking it, and his old workshop on the ground floor of number 11, under new occupancy now. He went into number 9, up to the second floor, and knocked on the door of Olly's apartment. And there she was, his dearest, truest friend, regarding him with utter astonishment, recovering her senses and welcoming him joyfully home.

There was only one thing missing, and it was quickly resolved. Gustav found the one person he had most longed to see, living alone in an apartment in the same building. His pride and delight, his beloved boy. Gustav threw his arms around Fritz and together they wept for joy.

They were home and together again.

Epilogue: Jewish Blood

Vienna, June 1954

An American GI stood looking across the Danube Canal toward Leopoldstadt. He wore dress uniform, with the chevron of a private first class on his sleeve. His unit patch was the 1st Infantry Division, whose troops had been among the first to hit Omaha Beach on D-Day. This soldier was far too young to have been there on that day: he'd been just a schoolboy in 1944. Now he was grown tall, the very image of a United States soldier. He was stationed in Bavaria, and had taken advantage of a one-week furlough to take a look at Vienna, the city where he'd been born.

It was familiar yet different—bringing itself back to life, healing its wounds. The GI approached the Soviet checkpoint and showed his identification. They waved him through, and he walked out across the broad Augarten Bridge, under the shadow of the Rossauer Kaserne, the grand imperial barracks where his parents had been married in 1917.

Many of the familiar buildings were scarred, some covered in scaffolding, still under repair. But Leopoldstadt was still recognizable, still as fresh in his mind as the day he'd left. How his life had changed since then, and how it had changed him. After high school, he had gone to college to study pharmacology, and in 1953 he'd been drafted into the army—Private Kurt Kleinmann. And now he was back.

Kurt was a product of America as much as Vienna now. His family were there—not only the Barnets, who had become family in all but name, but also Edith, who now lived in Connecticut. She and Richard had remained in London for three years after the war, but then finally left gloomy, impoverished England behind for good. The Paltenhoffers had adapted rapidly to American life. When they arrived, Peter and Joan—aged eight and six—had been English children with "Oxford accents" (according to the New Bedford paper), but that didn't last long. Determined to fit in, Richard and Edith changed their name from Paltenhoffer to Patten, and just this year, while Kurt was with the army overseas, they had become United States citizens.[1]

Strolling along the Obere Donaustrasse and up the Grosse Schiffgasse, Kurt was surprised at how well he remembered it all: the familiar turn right and left, and the Karmelitermarkt opened up before him, its stalls laid out in rows, the clock on its slender tower in the center, the shops and apartment buildings of Leopoldsgasse and Im Werd on either side. Just as it had always been.

As familiar as it all felt, he was an alien here now. The sense of foreignness was almost palpable—he couldn't even speak the language any more.

Kurt climbed the stairs and knocked on the door of Olga's apartment. The door was answered by his father. Gustav was older, more lined, with more gray in his hair, but still wearing the same familiar smile on the lean face with its trim moustache. And there was Olga herself, loyal, wonderful Olly. She was Frau Kleinmann now, Kurt's stepmother.

He visited many times during that summer. Sitting around the kitchen table, the four of them—Gustav and Olly, Kurt in his incongruous army uniform, and Fritz—talked as best they could. As time passed, Kurt found that a little of his German came back: just enough to get by, but not enough for a proper conversation.

It was hard to recover the lost years. His father didn't want to talk about his time in the camps, and Kurt's relationship with Fritz was altogether different. Raised as an all-American boy, Kurt was dismayed by his brother's communist sympathies. Fritz had acquired his politics by inheritance from their father's socialism, and in the camps from heroes like Robert Siewert and Stefan Heymann. Life as a worker in Soviet-controlled postwar Austria had confirmed him in his beliefs. There were also religious differences. None of the family aside from Kurt had ever been very devout, and Fritz had abandoned his faith entirely along the Auschwitz road.[2]

"No talking about politics or religion," decreed Gustav, and they stuck to safer subjects.

משפחה

On their return to Vienna in 1945, Gustav and Fritz had faced problems of adjustment. Even finding somewhere to live was a challenge in the bomb-damaged, Soviet-run city. Gustav stayed in Olly Steyskal's apartment, and remained there until he married her in 1948, the same year he managed to reestablish his upholstery business.

There was still anti-Semitism, but it had gone underground, confined to muttering and insinuation. Of the 183,000 Jews who had lived in Vienna in March 1938, more than two-thirds had emigrated: nearly 31,000 to Britain; 29,000 to the United States; 33,000 to South America, Asia, and Australia; and just over 9,000 to Palestine. Over 21,000 who had emigrated to European nations had subsequently come under Nazi rule, and nearly all went to the camps, along with 43,421 Jews deported directly from Vienna to Auschwitz, Łodz, Theresienstadt, and Minsk, and the thousands sent, like Fritz and Gustav, to Dachau and Buchenwald.[3]

Vienna after the Shoah still had a Jewish community, which gradually recovered its identity and preserved its heritage, but it was a fragment of what it had been. The synagogues had been destroyed or stood in ruins, and only a few were ever rebuilt. The Stadttempel in the ancient Jewish quarter, where Kurt had sung as a boy, was one of them.

Fritz was unable to work at first due to poor health, and lived for a time on a disability pension. He and his father discussed what they should do about Kurt. Should they bring him home? He was a child still, and they missed him. But what was there for him here? His mother was dead, and his father aging and poor. They concluded that he was better off where he was. So Gustav and Fritz carried on together, just the two of them, supporting each other as they had through so many trials.

One of the delights of those postwar years was their reunion with Alfred Wocher. The tough, courageous old German had survived the inferno of the last defense of the Reich, and tracked down his old Auschwitz friends in Vienna. He visited them many times. "For us concentration camp inmates, Wocher had fulfilled more than his duty," Fritz recalled. "Through his conduct he gave us courage and faith, and thus contributed decisively to our surviving Auschwitz. Nobody rewarded him for it. We the survivors are indebted to him."

While his father tried to forget what he had seen and suffered in the camps, Fritz was of an entirely different disposition. He remembered it vividly and deliberately and with anger. He harbored a burning detestation of the former Nazis who still lived in Vienna. He heard the older folk muttering about his papa—*See, the Kleinmann Jew is back again*—and while his father tried to live peacefully alongside the collaborators, Fritz wouldn't speak to anyone who had sided with Nazis. They were mystified by this, and one of the neighbors who had betrayed them to the SS actually complained to Gustav, "Your son won't say hello to us!" Willful ignorance about the Shoah was so entrenched that this man couldn't begin to grasp the evil of what he had done.

There were occasional reprisals against collaborators by younger Jewish men, and Fritz became involved. An Aryan neighbor, Sepp Leitner, had been a member of the Vienna-based 89th SS-Standarte, which had taken part in the destruction of the synagogues on Kristallnacht. Fritz confronted Leitner and beat him up. He was arrested for assault by the police, but the Soviet authorities, who approved of summary justice for fascists, ordered his release.

Fritz couldn't accept what his country had become; in Buchenwald he'd listened to the Austrian *Prominenten* debate the nation's post-Nazi future, imagining a democratic socialist utopia, and Fritz had longed for that. Things improved in 1955, when Austria regained its independence, but the workers' paradise never materialized. Fritz took evening classes and became active in his company's trade union. His family life was unsettled; he had two marriages, from which came a son, Peter, and a stepson, Ernst. When the state of Israel was founded, Fritz moved his family there, but even this failed to give him peace. Obliged to do military service in the Israeli army, Fritz found being at war too distressing to bear after his experiences in the concentration camps. After serving out his compulsory term of two years, he moved back to Vienna.

Meanwhile, Gustav was content to be back in his old trade and married to Olly. In 1964 he retired, having carried on to the ripe age of seventy-three. He and Olly visited America. Although he scarcely understood a word of English, he now had five American grandchildren and three great-grandchildren. He posed for photographs with the little ones on his knees, beaming contentedly, surrounded once more by love and family.

Gustav Kleinmann died on May 1, 1976, the day before his eighty-fifth birthday. He had been severely ill for some time, yet his prodigious inner strength had kept him going in his final days.

Two years later, Fritz, who was only in his mid-fifties, had to take early retirement. The torture he had endured in the Gestapo dungeon at Auschwitz had left him with permanent back injuries that, despite spinal operations, eventually caused partial paralysis. Nonetheless, he had his father's toughness, and he lived a long life, passing away on January 20, 2009, aged eighty-five.

✡

While his father tried to forget the Shoah, it was Fritz Kleinmann's abiding concern to ensure that the world did not. After the war ended, the Allies prosecuted high-ranking Nazis in trials at Nuremberg in 1945–46 and at Dachau in 1945–47. Many were executed or imprisoned, and the concepts of *genocide* and *crimes against humanity* entered the statutes of international law.

But once those trials were over, a shadow fell over Nazi atrocities—particularly within Germany itself. Those who had lived through it and colluded with it wanted to draw a veil over the past. By the end of the 1950s, a generation of Germans had been raised on a cushion of lies—that the Jews had mostly just emigrated, that there had been atrocities on all sides during the war, and that those committed by Germany had been no worse than those by the Allies. These young Germans knew almost nothing of the Holocaust, and the names of Auschwitz and Sobibor, Buchenwald and Belsen were obscure or unknown to them. Most of the SS murderers remained free, many still living in Germany.

That changed in 1963, when Fritz Bauer, a Jewish state

attorney in Frankfurt, who had helped trace Adolf Eichmann in Argentina, instituted proceedings against twenty-two former SS men accused of carrying out atrocities at Auschwitz. The witnesses in the Frankfurt trials included over two hundred surviving inmates, of whom ninety were Jews.[4] Among them were Gustav and Fritz Kleinmann, who gave written statements to the prosecutors in April and May 1963.[5] Their fellow witnesses included Stefan Heymann, Felix Rausch, and Gustav Herzog. Among those on trial were members of the camp Gestapo, Blockführers, and administrators. Some were acquitted; others received sentences ranging from three years to life.

More important than the individual sentences, the Frankfurt trials—along with Eichmann's trial in Jerusalem in 1961—forced Germany's eyes open, and ensured that the nation—and the world—would not forget the Holocaust.

Fritz Kleinmann continued doing his part. In 1987 he was invited by a friend, the Austrian political scientist Reinhold Gärtner, to give a public talk about his experiences to a group about to set out on a study trip to Auschwitz-Birkenau. Fritz would be one of four survivors speaking. "For days before it I could not sleep; the images from my concentration camp imprisonment welled up more intensely than ever before." The event—which included extracts from his father's diary read by a Viennese actor—moved Fritz profoundly, and overwhelmed the audience. He came back and gave his talk again and again to new parties for over a decade.

Persuaded to explore his memories further, Fritz wrote a short memoir that was later published in a book.[6] Even though decades had passed, he still burned with indignation and anger about the atrocities visited on him and his people, but his ire was countered by the love he still felt for those who had helped him survive: Robert Siewert, Stefan Heymann, Leo Moses, and all the rest. He pored over the handful of old documents he had preserved. He still had the photograph of himself taken in 1939 for his *J-Karte* identity card and the one taken in Buchenwald in 1940, which his mother had given to a relative before boarding the transport to Maly Trostinets.

And there was the diary. His father had revealed its existence to him shortly after their reunion in Vienna. Turning back its dog-eared cover, there was the first page, yellowed, covered with his father's angular pencil strokes, fading now after all these years. "Arrived in Buchenwald on the 2nd October 1939 . . ." The vividness of the images seared Fritz's mind. The quarry, hauling the stone-laden wagon up the tracks, the corpses in the mud, a man running across the sentry line and dropping with a bullet in his back, hanging from the beam in the Gestapo bunker, arms twisting out of their sockets, the weight of the Luger in his palm, the agonizing cold of the open wagon between Gleiwitz and Amstetten . . . and his father's poem, "Quarry Kaleidoscope," with its unforgettable central image:

> It rattles, the crusher, day out and day in,
> It rattles and rattles and breaks up the stone,
> Chews it to gravel and hour by hour
> Eats shovel by shovel in its guzzling maw.
> And those who feed it with toil and with care,
> They know it just eats, but will never be through.
> It first eats the stone and then eats them too.

But it hadn't crushed them all. A few, like the tall prisoner in the poem, had managed to outlast the machine, to keep going until the stone crusher clattered itself to a halt, malfunctioning, choked by its own appetite.

In the end the Kleinmann family had not only survived but prospered; through courage, love, solidarity, and blind luck, they outlasted the people who had tried to destroy them. They and their descendants spread and multiplied, perpetuating through the generations the love and unity that had helped them through the darkest of times. They took their past with them, understanding that the living must gather the memories of the dead and carry them into the safety of the future.

Bibliography and Sources

Interviews

Conducted by the author

Kurt Kleinmann: March–April 2016, July 2017
Peter Patten: April 2016, July 2017

Archived

Fritz Kleinmann: February 1997: interview 28129, Visual History Archive: University of Southern California Shoah Foundation Institute

Archive and unpublished sources

ABM   Archives of Auschwitz-Birkenau Museum, Oświęcim, Poland

AFB   Findbuch for Victims of National Socialism, Austria: findbuch.at (retrieved February 4, 2020)

AJJ   American Jewish Joint Distribution Committee Archives, New York

AWK   Testimonies from Kristallnacht: Wiener Library, London: available online at www.pogromnovem ber1938.co.uk/viewer/home (retrieved February 4, 2020)

BWM	Belohnungsakten des Weltkrieges 1914–1918: Mannschaftsbelohnungsanträge No. 45348, Box 21: Austrian State Archives, Vienna
DFK	Letters, photographs, and documents from the archive of Fritz Kleinmann
DKK	Letters and documents in possession of Kurt Kleinmann
DOW	Dokumentationsarchiv des Österreichischen Widerstandes, Vienna: some records available online at www.doew.at/personensuche (retrieved February 4, 2020)
DPP	Documents and photographs in possession of Peter Patten
DRG	Documents and photographs in possession of Reinhold Gärtner
FDR	FDR Presidential Library, Hyde Park, New York
FTD	Records of the Frankfurt Auschwitz trials: Fritz Bauer Institut, Frankfurt am Main, Germany
GRO	Records of births, marriages, and deaths for England and Wales: General Register Office, Southport, UK
HOI	Home Office: Aliens Department: Internees Index, 1939–1947: HO 396: National Archives, Kew, London
IKA	Archiv der Israelitischen Kultusgemeinde, Vienna
ITS	Documents on victims of Nazi persecution: ITS Digital Archive: International Tracing Service, Bad Arolsen, Germany
LJL	Leeds Jewish Refugee Committee: case files: WYL 5044/12: West Yorkshire Archive Service, Leeds, UK
LJW	Leeds Jewish Refugee Committee: correspondence and papers: Collection 599: Wiener Library, London
MTW	Maly Trostinec witness correspondence, 1962–67: World Jewish Congress Collection: Box C213-05: American Jewish Archives, Cincinnati

NARA   National Archives and Records Administration, Washington DC

PGB   Prisoner record archive: KZ-Gedenkstätte Buchenwald, Weimar

PGD   Prisoner record archive: KZ-Gedenkstätte Dachau, Dachau

PGM   Prisoner record archive: KZ-Gedenkstätte Mauthausen Research Center, Vienna

PNY   Passenger Lists of Vessels Arriving at New York: Microfilm Publication M237, 675: NARA

TAE   Trial of Adolf Eichmann: District Court Sessions: State of Israel Ministry of Justice: available online at nizkor.org (retrieved February 4, 2020)

WLO   *Adolph Lehmanns Adressbuch*: Wienbibliothek Digital: www.digital.wienbibliothek.at/wbrobv/periodical/titleinfo/5311 (retrieved February 4, 2020)

YVP   Papers and documents: Yad Vashem, Jerusalem: some available online at www.yadvashem.org

YVS   Central Database of Shoah Victims' Names: Yad Vashem, Jerusalem: available online at yvng.yadvashem.org (retrieved February 4, 2020)

Books and articles

Aarons, Mark. *War Criminals Welcome: Australia, a Sanctuary for Fugitive War Criminals Since 1945.* Melbourne: Black Inc., 2001.

Arad, Yitzhak, Israel Gutman, and Abraham Margaliot. *Documents on the Holocaust*, 8th ed. Translated by Lea Ben Dor. Lincoln, NE, and Jerusalem: University of Nebraska Press and Yad Vashem, 1999.

Bardgett, Suzanne and David Cesarani, eds. *Belsen 1945: New Historical Perspectives.* London: Vallentine Mitchell, 2006.

Barton, Waltraud, ed. *Ermordet in Maly Trostinec: die österreichischen Opfer der Shoa in Weißrussland.* Vienna: New Academic Press, 2012.

Bentwich, Norman. "The Destruction of the Jewish Community in Austria 1938–1942," in *The Jews of Austria,* edited by Josef Fraenkel, pp. 467–78. London: Vallentine Mitchell, 1970.

Berkley, George E. *Vienna and Its Jews: The Tragedy of Success, 1880s–1980s.* Cambridge, MA: Abt Books, 1988.

Browning, Christophe. *The Origins of the Final Solution.* London: Arrow, 2005.

Burkitt, Nicholas Mark. *British Society and the Jews.* University of Exeter: PhD dissertation, 2011.

Cesarani, David. *Eichmann: His Life and Crimes.* London: Vintage, 2005.

———. *Final Solution: The Fate of the Jews 1933–49.* London: Macmillan, 2016.

Czech, Danuta. *Auschwitz Chronicle: 1939–1945.* London: I. B. Tauris, 1990.

Czeike, Felix. *Historisches Lexikon Wien,* 6 vols. Vienna: Kremayr & Scheriau, 1992–97.

Długoborski, Wacław and Franciszek Piper, eds. *Auschwitz 1940–1945: Studien der Geschichte des Konzentrations- und Vernichtungslagers Auschwitz,* 5 vols. Oświęcim: Verlag des Staatlichen Museums Auschwitz-Birkenau, 1999.

Dobosiewicz, Stanisław. *Mauthausen-Gusen: obóz zagłady.* Warsaw: Wydawn, 1977.

Dror, Michael. "News from the Archives." *Yad Vashem Jerusalem* 81 (October 2016): p. 22.

Dutch, Oswald. *Thus Died Austria.* London: E. Arnold, 1938.

Fein, Erich and Karl Flanner. *Rot-Weiss-Rot in Buchenwald.* Vienna: Europaverlag, 1987.

Foreign Office (UK). *Papers Concerning the Treatment of German Nationals in Germany, 1938–1939.* London: HMSO, 1939.

Friedländer, Saul. *Nazi Germany and the Jews: vol. 1: The Years of Persecution, 1933–1939*. London: Weidenfeld and Nicolson, 1997.

Friedman, Saul S. *No Haven for the Oppressed: United States Policy Toward Jewish Refugees, 1938–1945*. Detroit: Wayne State University Press, 1973.

Frieser, Karl-Heinz. *The Eastern Front, 1943–1944*. Translated by Barry Smerin and Barbara Wilson. Oxford: Clarendon Press, 2017.

Gärtner, Reinhold and Fritz Kleinmann. *Doch der Hund will nicht krepieren: Tagebuchnotizen aus Auschwitz*. Innsbruck: Innsbruck University Press, 1995, 2012.

Gedye, G. E. R. *Fallen Bastions: The Central European Tragedy*. London: Gollancz, 1939.

Gemeinesames Zentralnachweisbureau. *Nachrichten über Verwundete und Kranke Nr 190 ausgegeben am 6.1.1915; Nr 203 ausgegeben am 11.1.1915*. Vienna: k. k. Hof und Staatsdockerei, 1915.

Gerhardt, Uta and Thomas Karlauf, eds. *The Night of Broken Glass: Eyewitness Accounts of Kristallnacht*. Translated by Robert Simmons and Nick Somers. Cambridge: Polity Press, 2012.

Gerlach, Christian. *Kalkulierte Morde: Die deutsche Wirtschafts- und Vernichtungspolitik in Weißrußland 1941 bis 1944*. Hamburg: Hamburger Edition, 1999.

Gilbert, Martin. *Auschwitz and the Allies*. London: Michael Joseph, 1981.

———. *The Holocaust: The Jewish Tragedy*. London: Collins, 1986.

———. *The Routledge Atlas of the Holocaust*, 3rd ed. London: Routledge, 2002.

Gillman, Peter and Leni Gillman. *"Collar the Lot!": How Britain Interned and Expelled Its Wartime Refugees*. London: Quartet, 1980).

Gold, Hugo. *Geschichte der Juden in Wien: Ein Gedenkbuch*. Tel Aviv: Olamenu, 1966.

Goltman, Pierre. *Six mois en enfer*. Paris: Éditions le Manuscrit, 2011.

Gottwaldt, Alfred and Diana Schulle. *Die «Judendeportationen» aus dem Deutschen Reich 1941–1945*. Wiesbaden: Marix Verlag, 2005.

Grenville, Anthony. "Anglo-Jewry and the Jewish Refugees from Nazism." *Association of Jewish Refugees Journal* (December 2012); available online at ajr.org.uk/wp-content/uploads/2018/02/2002 _December.pdf (retrieved February 4, 2020).

Gutman, Yisrael and Michael Berenbaum, eds. *Anatomy of the Auschwitz Death Camp.* Bloomington, IN: Indiana University Press, 1994.

Hackett, David A., ed., transl. *The Buchenwald Report.* Boulder, CO: Westview Press, 1995.

Haunschmied, Rudolf A., Jan-Ruth Mills, and Siegi Witzany-Durda. *St. Georgen-Gusen-Mauthausen: Concentration Camp Mauthausen Reconsidered.* Norderstedt: Books on Demand, 2007.

Hayes, Peter. *Industry and Ideology: IG Farben in the Nazi Era.* Cambridge: Cambridge University Press, 2001.

Hecht, Dieter J. " 'Der König rief, und alle, alle kamen': Jewish Military Chaplains on Duty in the Austro-Hungarian Army During World War I." *Jewish Culture and History* 17/3 (2016): pp. 203–16.

Heimann-Jelinek, Felicitas, Lothar Höbling and Ingo Zechner, *Ordnung muss sein: Das Archiv der Israelitischen Kultusgemeinde Wien.* Vienna: Jüdisches Museum Wien, 2007.

Heller, Peter. "Preface to a Diary on the Internment of Refugees in England in the Year of 1940" in *Exile and Displacement.* Edited by. Lauren Levine Enzie. New York: Peter Lang, 2001.

Horsky, Monika. *Man muß darüber reden. Schüler fragen KZ-Häftlinge.* Vienna: Ephelant Verlag, 1988.

Jones, Nigel. *Countdown to Valkyrie: The July Plot to Assassinate Hitler.* London: Frontline, 2008.

Keegan, John. *The First World War.* London: Hutchinson, 1998.

Kershaw, Roger. "Collar the lot! Britain's Policy of Internment During the Second World War," UK National Archives Blog, July 2, 2015, blog.nationalarchives.gov.uk/blog/collar-lot-britains-policy -internment-second-world-war (retrieved February 4, 2020).

K.u.k. Kriegsministerium. *Verlustliste Nr 209 ausgegeben am 13.7.1915.* Vienna: k. k. Hof und Staatsdockerei, 1915.

——. *Verlustliste Nr 244 ausgegeben am 21.8.1915.* Vienna: k. k. Hof und Staatsdockerei, 1915.

Kurzweil, Edith. *Nazi Laws and Jewish Lives.* London: Transaction, 2004.

Langbein, Hermann. *Against All Hope: Resistance in the Nazi Concentration Camps, 1938–1945.* Translated by Harry Zohn. London: Constable, 1994.

——. *People in Auschwitz.* Translated by Harry Zohn. Chapel Hill, NC: University of North Carolina Press, 2004.

Le Chêne, Evelyn. *Mauthausen: The History of a Death Camp.* Bath, UK: Chivers, 1971.

Levi, Primo. *Survival in Auschwitz and The Reawakening: Two Memoirs.* New York: Summit, 1986; previously publ. 1960, 1965.

Loewenberg, Peter. "The *Kristallnacht* as a Public Degradation Ritual" in *The Origins of the Holocaust.* Edited by Michael Marrus. London: Meckler, 1989.

London, Louise. *Whitehall and the Jews, 1933–1948: British Immigration Policy, Jewish Refugees and the Holocaust.* Cambridge: Cambridge University Press, 2000.

Lowenthal, Marvin. *The Jews of Germany.* London: L. Drummond, 1939.

Lucas, James. *Fighting Troops of the Austro-Hungarian Army, 1868–1914.* Tunbridge Wells, UK: Spellmount, 1987.

Maier, Ruth. *Ruth Maier's Diary: A Young Girl's Life under Nazism.* Translated by Jamie Bulloch. London: Harvill Secker, 2009.

Mazzenga, Maria, ed. *American Religious Responses to Kristallnacht.* New York: Palgrave Macmillan, 2009.

Megargee, Geoffrey P., ed. *The United States Holocaust Memorial Museum Encyclopedia of Camps and Ghettos, 1933–1945,* 4 vols. Bloomington, IN: Indiana University Press, 2009.

Pendas, Devin O. *The Frankfurt Auschwitz Trial, 1963–1965.* Cambridge: Cambridge University Press, 2006.

Phillips, Raymond. *Trial of Josef Kramer and Forty-Four Others: The Belsen Trial.* London: W. Hodge, 1949.

Plänkers, Tomas. *Ernst Federn: Vertreibung und Rückkehr. Interviews zur Geschichte Ernst Federns und der Psychoanalyse.* Tübingen: Diskord, 1994.

Pukrop, Marco. "Die SS-Karrieren von Dr. Wilhelm Berndt und Dr. Walter Döhrn. Ein Beitrag zu den unbekannten KZ-Ärzten der Vorkriegszeit." *Werkstatt Geschichte* 62 (2012): pp. 76–93.

Rabinovici, Doron. *Eichmann's Jews: The Jewish Administration of Holocaust Vienna, 1938–1945.* Translated by Nick Somers. Cambridge: Polity Press, 2011.

Rees, Laurence. *The Holocaust: A New History.* London: Viking, 2017.

Rosenkranz, Herbert. "The Anschluss and the Tragedy of Austrian Jewry 1938–1945" in *The Jews of Austria.* Edited by Josef Fraenkel. London: Valentine, Mitchell, 1970.

Sagel-Grande, Irene, H. H. Fuchs, and C. F. Rüter. *Justiz und NS-Verbrechen: Sammlung Deutscher Strafurteile wegen Nationalsozialistischer Tötungsverbrechen 1945–1966: Band XIX.* Amsterdam: University Press Amsterdam, 1978.

Schindler, John R. *Fall of the Double Eagle: The Battle for Galicia and the Demise of Austria–Hungary.* Lincoln, NE: University of Nebraska Press, 2015.

Silverman, Jerry. *The Undying Flame: Ballads and Songs of the Holocaust.* Syracuse, NY: Syracuse University Press, 2002.

Sington, Derrick. *Belsen Uncovered.* London: Duckworth, 1946.

Stein, Harry (compiler). *Buchenwald Concentration Camp 1937–1945.* Edited by Gedenkstätte Buchenwald. Göttingen: Wallstein Verlag, 2004.

Taylor, Melissa Jane. *"Experts in Misery?" American Consuls in Austria, Jewish Refugees and Restrictionist Immigration Policy, 1938–1941.* University of South Carolina: PhD dissertation, 2006.

Teichova, Alice. "Banking in Austria" in *Handbook on the History of European Banks*. Edited by Manfred Pohl. Aldershot: Edward Elgar, 1994.

Trimble, Lee with Jeremy Dronfield. *Beyond the Call*. New York: Berkley, 2015.

van Pelt, Robert Jan. *The Case for Auschwitz: Evidence from the Irving Trial*. Bloomington, IN: Indiana University Press, 2016.

van Pelt, Robert Jan and Debórah Dwork. *Auschwitz: 1270 to the Present*. New Haven, CT: Yale University Press, 1996.

Wachsmann, Nikolaus. *KL: A History of the Nazi Concentration Camps*. London: Little, Brown, 2015.

Wagner, Bernd C. *IG Auschwitz: Zwangsarbeit und Vernichtung von Häftlingen des Lagers Monowitz 1941–1945*. Munich: K. G. Saur, 2000.

Wallner, Peter. *By Order of the Gestapo: A Record of Life in Dachau and Buchenwald Concentration Camps*. London: John Murray, 1941.

Walter, John. *Luger: The Story of the World's Most Famous Handgun*. Ebook ed. Stroud, UK: History Press, 2016.

Wasserstein, Bernard. *Britain and the Jews of Europe, 1939–1945*. London: Leicester University Press, 1999.

Watson, Alexander. *Ring of Steel: Germany and Austria-Hungary at War, 1914–1918*. London: Penguin, 2014.

Weinzierl, Erika. "Christen und Juden nach der NS-Machtergreifung in Österreich" in *Anschluß 1938*. Vienna: Verlag für Geschichte und Politik, 1981.

Werber, Jack and William B. Helmreich. *Saving Children*. London: Transaction, 1996.

Wünschmann, Kim. *Before Auschwitz: Jewish Prisoners in the Prewar Concentration Camps*. Cambridge, MA: Harvard University Press, 2015.

Wyman, David S. *America and the Holocaust*, 13 vols. London: Garland, 1990.

Zalewski, Andrew. *Galician Portraits: In Search of Jewish Roots.* Jenkintown, PA: Thelzo Press, 2012.

Zucker, Bat-Ami. *In Search of Refuge: Jews and US Consuls in Nazi Germany, 1933–1941.* London: Vallentine Mitchell, 2001.

Acknowledgments

This book could not have been written without its primary source material—Gustav Kleinmann's concentration camp diary and Fritz Kleinmann's memoir, which came to me through Professor Reinhold Gärtner of Innsbruck University. Reinhold helped Fritz to publish both documents in the book *Doch der Hund will nicht krepieren* ("But still the dog will not die": Innsbruck University Press, 2012), and has provided indispensable cooperation in my initial research for this book, for which I thank him.

I am profoundly grateful to Kurt Kleinmann, who lived through the Anschluss and the Nazi occupation of Vienna, for the many hours of interviews and months of correspondence. Without Kurt's generous and tireless help, this tale would have been far less rich in depth and detail. Peter Patten, Gustav's grandson, has also very kindly contributed interviews and correspondence. I am grateful also to Rachel Schine, who helped put me in touch with the American branch of the family. The Austrian side of the family have also provided support. The encouragement of Peter Kleinmann, Victor Zehetbauer, and his father, Ernst, as well as Richard Wilczek, has been indispensable.

A draft English translation of *Doch der Hund* prepared by John Rie was my first contact with this story, and provided a vital foundation for creating my own translation of Gustav's diary and Fritz's memoir. For the preparation of the Hebrew section titles, I am grateful for the expert advice given by Keren Joseph-Browning. Meticulous copyediting by Richenda Todd has saved me from many small but awkward blunders. Júlia Moldova, my Hungarian translator, drew my attention to an important historical omission in the original edition of the book, corrected herein.

Many archives and their archivists have provided me with guidance, documents, and images, and have patiently dealt with my queries. I am grateful to all of them. They include the Austrian State Archive, Vienna, for documents on Gustav Kleinmann's First World War record; Douglas Ballman and Georgiana Gomez, Access Supervisor, University of Southern California Shoah Foundation Institute for Visual History and Education, for providing a transcript of Fritz Kleinmann's 1997 interview and helping with photographs; Ewa Bazan, Head of the Bureau for Former Prisoners at Auschwitz-Birkenau Memorial Museum; Johannes Beermann, archivist, Fritz Bauer Institut, Goethe-Universität, Frankfurt am Main, for Fritz and Gustav's witness statements from the Frankfurt Auschwitz trials; Cambridge University Library; Judy Farrar, Archives and Special Collections Librarian, Claire T. Carney Library, University of Massachusetts, Dartmouth, for information on Samuel Barnet; Harriet Harmer, Archive Assistant, West Yorkshire Archive Service, Leeds, UK, for documents on Edith Kleinmann and Richard Paltenhoffer; Elisa Ho, Archivist and Special Projects Coordinator, the Jacob Rader Marcus Center of the American Jewish Archives, Cincinnati, for documents on Maly Trostinets; Katharina Kniefacz, KZ-Gedenkstätte Mauthausen Research Center, Vienna, for prisoner records on Fritz Kleinmann; Albert Knoll, archivist, KZ-Gedenkstätte Dachau, for information on Richard Paltenhoffer; Kimberly Kwan, volunteer, Gedenkstätte Buchenwald, for information on the Kleinmanns and Richard Paltenhoffer; Heike Müller, International Tracing Service, Bad Arolsen, Germany, for documents relating to the Kleinmanns' incarceration in various concentration camps; Susanne Uslu-Pauer,

Head of Department, Archive of the Israelitischen Kultusgemeinde, Vienna; and the Wiener Library, London.

Finally, I am grateful to my literary agent, Andrew Lownie, for first bringing the Kleinmann story to my attention, to Yuval Taylor for giving me the chance to make it real, and to Dan Bunyard and Zennor Compton at Penguin Books UK, for believing in the book and bringing their enthusiasm to the project. My thanks likewise go to Sara Nelson and Doug Jones at HarperCollins US. As ever, my partner, Kate, has provided the constant, patient moral support that has sustained me through every book I have ever written.

Jeremy Dronfield, February 2020

Notes

Prologue

1 Moon phase data from www.timeanddate.com/moon/austria /amstetten?month=1&year=1945.

Chapter 1: "When Jewish Blood Drips from the Knife..."

1 Printed in *Die Stimme*, March 11, 1938, p. 1; see also G. E. R. Gedye, *Fallen Bastions: The Central European Tragedy* (1939), pp. 287–9, for an eyewitness account of events in Vienna that day.

2 Schuschnigg's Fatherland Front Party was fascistic, suppressing the Nazi Party and the Social Democrats. However, it was not especially anti-Semitic. On the number of Jews in Austria see Martin Gilbert, *The Routledge Atlas of the Holocaust* (2002), p. 22, and Norman Bentwich, "The Destruction of the Jewish Community in Austria 1938–1942" in *The Jews of Austria*, ed. Josef Fraenkel (1970), p. 467.

3 *Die Stimme*, March 11, 1938, p. 1.

4 Some people of Jewish descent considered themselves entirely German; Peter Wallner, a Viennese, stated, "Nor was I ever a Jew, though all four of my grandparents were Jewish." But when the Nazis came he was persecuted along with the rest; "For according to the Nuremberg Laws I am a Jew" (Peter Wallner, *By Order of the Gestapo: A Record of Life in Dachau and Buchenwald Concentration Camps* [1941], pp. 17–18). Under the Nuremberg Laws of 1935 a person was defined as Jewish, regardless of religion, if they had more than two Jewish grandparents.

5 *Die Stimme*, March 11, 1938, p. 1.

6 *Jüdische Presse*, March 11, 1938, p. 1.

7 The scenes on this day are described by George Gedye (*Fallen Bastions*, pp. 287–96), a British journalist for the *Daily Telegraph* and *New York Times*, who lived in Vienna.

8 For this reason, Schuschnigg had cynically set the minimum age for voting in the plebiscite at twenty-four; most Nazis were below that age.

9 *The Times*, March 11, 1938, p. 14; also *Neues Wiener Tagblatt (Tages-Ausgabe)*, March 11, 1938, p. 1.

10 Gedye (*Fallen Bastions*, pp. 290–93) describes the scenes as the evening progressed.

11 Ibid., p. 290; *The Times*, March 12, 1938, p. 12.

12 Quoted in Gedye, *Fallen Bastions*, pp. 10, 293, and *The Times*, March 12, 1938, p. 12. According to *The Times*, newspapers in Berlin that evening claimed that Germany had quashed "treason" by the "Marxist rats" in the Austrian government who had been carrying out "harrowing cruelties" against the people, who were fleeing to the German border in large numbers. With these untruths the Nazis justified their move to take over Austria.

13 The synagogue that evening is described as "*überfüllt*"—"overcrowded," "jam-packed" (Hugo Gold, *Geschichte der Juden in Wien: Ein Gedenkbuch* (1966), p. 77; Erika Weinzierl, "Christen und Juden nach der NS-Machtergreifung in Österreich" in *Anschluß 1938* (1981): pp. 197–98).

14 Gedye, *Fallen Bastions*, p. 295. The hostility to Catholics stemmed from antagonism over issues such as Nazi attempts to suppress the Old Testament and de-Judaize Christianity, as well as the Church's recognition of non-Aryan Christian converts and the Vatican's condemnation of racism (David Cesarani, *Final Solution: The Fate of the Jews 1933–49* (2016), pp. 114–16, 136).

15 Quoted in Cesarani, *Final Solution*, p. 148.

16 Gedye, *Fallen Bastions*, p. 295.

17 Oswald Dutch, *Thus Died Austria* (1938), pp. 231–32; see also *Neues Wiener Tagblatt*, March 12, 1938, p. 3; *Banater Deutsche Zeitung*, March 13, 1938, p. 5; *The Times*, March 14, 1938, p. 14.

18 *Neues Wiener Tagblatt*, March 12, 1938, p. 3.

19 Gedye, *Fallen Bastions*, p. 282.

20 *Arbeitersturm*, March 13, 1938, p. 5; *The Times*, April 17, 1938, p. 14.

21 It isn't certain which police station this was. The most likely is Leopoldsgasse, a station of the Schutzpolizei Gruppenkommando Ost, the uniformed Reich police (*Reichsamter und Reichsbehörden in der Ostmark*, p. 207, AFB).

22 Based on Fritz Kleinmann's memoir: Reinhold Gärtner and Fritz Kleinmann, *Doch der Hund will nicht krepieren: Tagebuchnotizen aus Auschwitz* (2012); also testimony of Kurt Kleinmann and Edith's son, Peter Patten; additional details from various contemporary sources.

23 Evidence of Moritz Fleischmann, vol. 1, session 17, TAE; George E. Berkley, *Vienna and Its Jews: The Tragedy of Success, 1880s–1980s* (1988), p. 259; Marvin Lowenthal, *The Jews of Germany* (1939), p. 430. See also *The Times*, March 31, 1938, p. 13; April 7, 1938, p. 13.

24 Gedye, *Fallen Bastions*, p. 354.

25 *The Times*, April 8, 1938, p. 12; April 11, 1938, p. 11; also Gedye, *Fallen Bastions*, p. 9.

26 *The Times*, April 11, 1938, p. 12. Even the ballot paper itself was a work of propaganda, with a big circle in the center for "yes" (to the Anschluss) and a little one off to the side for "no."

27 *The Times*, April 12, 1938, p. 14.

28 *The Times*, April 9, 1938, p. 11.

29 *The Times*, March 23, 1938, p. 13; March 26, 1938, p. 11; April 30, 1938, p. 11.

30 Bentwich, "Destruction." p. 470.

31 Ibid.; Herbert Rosenkranz, "The Anschluss and the Tragedy of Austrian Jewry 1938–1945" in *The Jews of Austria*, ed. Josef Fraenkel (1970), p. 484.

32 Dachau, established in 1933 in a disused factory, was the first dedicated concentration camp. By summer 1938 there were four major operational camps in Germany (plus some smaller ones): Dachau, Buchenwald, Sachsenhausen, and Flossenbürg, with several more opening shortly after—including Mauthausen in Austria, which opened in August 1938 (see Nikolaus Wachsmann, *KL: A History of the Nazi Concentration Camps* [2015]; Cesarani, *Final Solution*; Laurence Rees, *The Holocaust: A New History* [2017].

33 Reich Ministry of the Interior regulations, August 17, 1938, in Yitzhak Arad, Israel Gutman and Abraham Margaliot, *Documents on the Holocaust*, 8th ed., trans. Lea Ben Dor (1999), pp. 98–99.

34 Testimony B.306, AWK.

35 Testimony B.95, AWK.

36 This was the story according to the Brussels correspondent for *The Times* (October 27, 1938, p. 13). Associated Press via the *Chicago Tribune* (October 27, 1938, p. 15) added the detail about the camera, increased the number of Nazis involved to four, and added the anonymous claim that the Nazis had been knocked down and kicked.

37 *Neues Wiener Tagblatt*, October 26, 1938, p. 1.

38 *Völkischer Beobachter*, October 26, 1938, p. 1, quoted in Peter Loewenberg, "The *Kristallnacht* as a Public Degradation Ritual" in *The Origins of the Holocaust*, ed. Michael Marrus (1989), p. 585.

39 *Neues Wiener Tagblatt*, November 8, 1938, p. 1.

40 "Night of broken glass" is the usual translation, but "crystal" is more accurate.

41 Telegram from Reinhard Heydrich to all police headquarters, November 10, 1938, in Arad et al., *Documents*, pp. 102–4.

42 UK consul-general in Vienna, letter, November 11, 1938, in Foreign Office (UK), *Papers Concerning the Treatment of German Nationals in Germany 1938–1939* (1939), p. 16.

Chapter 2: Traitors to the People

1 The Polizeiamt Leopoldstadt, headquarters of the local uniformed police, was at Ausstellungstrasse 171 (*Reichsamter und Reichsbehörden in der Ostmark*, p. 204, AFB).

2 Narrative based on the memoir of Fritz Kleinmann in Gärtner and Kleinmann, *Doch der Hund*, p. 188; additional details: witness testimonies B.24 (anon.), B.62 (Alfred Schechter), B.143 (Carl Löwenstein), AWK; also testimonies of Siegfried Merecki (Manuscript 166 [156], Margarete Neff (Manuscript 93 [205] in Uta Gerhardt and Thomas Karlauf, eds. *The Night of Broken Glass: Eyewitness Accounts of Kristallnacht*, trans. Robert Simmons and Nick Somers (2012); Wallner, *By Order of the Gestapo*.

3 UK consul-general in Vienna, letter, November 11, 1938, in Foreign Office (UK), *Papers Concerning the Treatment of German Nationals in Germany, 1938–1939* (1939), p. 16.

4 The exact number of documented arrests is 6,547 (Melissa Jane Taylor, "*Experts in Misery*"? *American Consuls in Austria, Jewish Refugees and Restrictionist Immigration Policy, 1938–1941* [2006], p. 48).

5 B.62 (Alfred Schechter), AWK. At this time, Mauthausen camp was for convicts; Jews were not imprisoned there prior to the war, but it was believed at the time that they were (e.g., *The Scotsman*, November 14, 1938; cf. Kim Wünschmann, *Before Auschwitz: Jewish Prisoners in the Prewar Concentration Camps* [2015], p. 183).

6 B.143 (Carl Löwenstein), AWK.

7 *New York Times*, November 15, 26, 1938, p. 1.

8 Quoted in Swiss *National Zeitung*, November 16, 1938.

9 *The Spectator*, November 18, 1938, p. 836.

10 *Westdeutscher Beobachter* (Cologne), November 11, 1938.

11 Ibid.

12 Unnamed German newspaper, quoted by UK consul-general in Vienna, November 11, 1938, in Foreign Office, *Papers*, p. 15.

13 Cesarani, *Final Solution*, p. 199.

14 *The Spectator*, November 18, 1938, p. 836.

15 David Cesarani, *Eichmann: His Life and Crimes* (2005), pp. 60ff.

16 Heydrich, quoted in Cesarani, *Final Solution*, p. 207.

17 Doron Rabinovici, *Eichmann's Jews: The Jewish Administration of Holocaust Vienna, 1938–1945*, trans. Nick Somers (2011), pp. 50ff.; Cesarani, *Final Solution*, pp. 147ff.

18 *The Spectator*, July 29, 1938, p. 189.

19 Ibid., August 19, 1938, p. 294.

20 Adolf Hitler, speech to the Reichstag, January 30, 1939, quoted in *The Times*, January 31, 1939, p. 14; also in Arad et al., *Documents*, p. 132.

21 *Daily Telegraph*, November 22, 1938; also House of Commons *Hansard*, November 21, 1938, vol. 341, cc1428–83.

22 Ibid.

23 Testimony B.226, AWK.

24 *The Times*, December 3–12, 1938.

25 Fritz Kleinmann, 1997 interview.

26 *Manchester Guardian*, December 15, 1938, p. 11; March 18, 1939, p. 18.

27 Letter from Leeds JRC to Overseas Settlement Dept, JRC, London, June 7, 1940, LJL.

28 *The Times*, classified ads, 1938–39 *passim*.

29 Louise London, *Whitehall and the Jews, 1933–1948: British Immigration Policy, Jewish Refugees and the Holocaust* (2000), p. 79.

30 *The Times*, November 8, 1938, p. 4.

31 The system could only cope with investigating a limited number of applicants; women applying to be servants were easier to vet than men, and so over half of the Jews entering Britain in 1938–39 were women (Cesarani, *Final Solution*, p. 158). The Home Office expedited the process by having Jewish refugee agencies process the applications, which increased the rate to four hundred a week (ibid., p. 214).

32 Letter from UK consul-general in Vienna, November 11, 1938, in Foreign Office, *Papers*, p. 15.

33 This building, at Wallnerstrasse 8, now houses the Vienna Stock Exchange.

34 M. Mitzmann, "A Visit To Germany, Austria and Poland in 1939," document 0.2/151, YVP.

35 Harry Stein (compiler), *Buchenwald Concentration Camp 1937–1945*, ed. Gedenkstätte Buchenwald (2004), pp. 115–16; Gärtner and Kleinmann, *Doch der Hund*, pp. 80–81.

36 Fritz recalled (1997 interview) that the third man was called Schwarz, although no record has been found of a person of that name living in Im Werd 11. Fritz was unable to recall the name of the fourth member of the group (the building's Nazi leader).

37 The dialogue here is from interviews given by Fritz and Kurt Kleinmann. They both recalled these scenes quite vividly.

38 Buchenwald prisoner record card 1.1.5.3/6283389, ITS.

Chapter 3: Blood and Stone

1 This account is based primarily on Gustav Kleinmann's diary and Fritz's recollections, with additional circumstantial details from other sources (e.g., Jack Werber and William B. Helmreich, *Saving Children* [1996], pp. 1–3, 32–6; Stein, *Buchenwald*, pp. 115–16; testimonies B.82, B.192, B.203, AWK).

2 Fritz Kleinmann (in Gärtner and Kleinmann, *Doch der Hund*, p. 12) gives a figure of 1,048 Viennese Jews in this transport, but other sources (Stein, *Buchenwald*, p. 116) give 1,035.

3 Stein, *Buchenwald*, pp. 27–28.

4 See e.g., testimony B.203, AWK.

5 Gärtner and Kleinmann, *Doch der Hund*, p. 15n.

6 Stein, *Buchenwald*, p. 35.

7 Buchenwald prisoner record cards 1.1.5.3/6283376, 1.1.5.3 /6283389, ITS. There were no tattoos; this practice was begun at Auschwitz in November 1941 and was not employed at any other camp (Wachsmann, *KL*, p. 284).

8 Werber and Helmreich, *Saving Children*, p. 36.

9 Testimony B.192, AWK.

10 The basic concentration camp badge was an inverted triangle, the color of which denoted a category: red for political prisoners, green for criminals, pink for homosexuals, and so on. For Jewish prisoners the category badge was combined with a second, yellow, triangle, making up a Star of David; if the Jewish prisoner didn't fit into any of the other categories, both triangles were yellow.

11 Emil Carlebach, in David A. Hackett, ed., trans. *The Buchenwald Report* (1995), pp. 162–3.

12 This is not the same as the "little camp" set up in 1943 to the north of the barracks (Stein, *Buchenwald*, pp. 149–51). There is a detailed description of the original little camp in 1939–40 by inmate Felix Rausch in Hackett, *Buchenwald Report*, pp. 271–76.

13 Hackett, *Buchenwald Report*, pp. 113–14. Following Kristallnacht, new arrivals totalled 10,098. There were over 9,000 subsequent departures due to release, transfer, or death (about 2,000 deaths in total in 1938–39, not including those who died between Weimar and the camp; ibid., p. 109). The prisoner population of Buchenwald declined steeply from 1938–39, exploding again with the autumn 1939 intake (8,707 during September–October).

14 Fritz wrote later: "I know that my father risked his life with this diary. None of the other prisoners had encouraged him to do this, as he was putting not only himself but all of us at risk. And even today, questions remain unanswered: Where did my father hide the diary? How did he get it through the controls?" (Gärtner and Kleinmann, *Doch der Hund*, pp. 12–13). Gustav did reveal that at one point, when he was a room orderly in his barrack, he hid it inside the bunks, and when he was on outdoor work he carried it on his person (Fritz Kleinmann, 1997 interview).

15 This account is based primarily on Gustav Kleinmann's diary and Fritz's recollections, with additional circumstantial details

from other sources (e.g., Hackett, *Buchenwald Report*; Stein, *Buchenwald*; testimony B.192, AWK).

16 In Himmler's words, the kapo's task was "to see that the work gets done . . . As soon as we are no longer satisfied with him, he is no longer a kapo and returns to the other inmates. He knows that they will beat him to death his first night back" (quoted in Rees, *Holocaust*, p. 79).

17 Based on size of wagon and density of broken limestone = 1,554 kg/m^3. Different sources give the number of men assigned to pull each wagon as between sixteen and twenty-six.

18 Gustav refers to this place as the *"Todes-Holzbaracke"* ("death barrack"), probably a nickname for a building used for sick Jews after they were barred from the prisoners' infirmary (block 2, in the southwest corner of the camp facing onto the roll-call square) in September 1939 (see Emil Carlebach in Hackett, *Buchenwald Report*, p. 162).

19 Stein, *Buchenwald*, p. 96.

20 Stefan Heymann in Hackett, *Buchenwald Report*, p. 253.

21 Nigel Jones, *Countdown to Valkyrie: The July Plot to Assassinate Hitler* (2008), pp. 103–5.

22 Wachsmann, *KL*, p. 220.

23 Hackett, *Buchenwald Report*, p. 51; Stein, *Buchenwald*, p. 119.

24 Hackett, *Buchenwald Report*, pp. 231, 252–53; Wachsmann, *KL*, p. 220.

25 Fritz Kleinmann, quoted in Monika Horsky, *Man muß darüber reden. Schüler fragen KZ-Häftlinge* (1988), pp. 48–49, reproduced in Gärtner and Kleinmann, *Doch der Hund*, p. 16n.

26 Stein, *Buchenwald*, pp. 52, 108–9; testimony B.192, AWK.

27 Heller later served as a doctor in Auschwitz. He survived the Holocaust and emigrated to the United States. "He was a very decent man. If he could help a person, he would," recalled one of his fellow prisoners (obituary, *Chicago Tribune*, September 29, 2001).

28 Hackett, *Buchenwald Report*, pp. 60–64.

29 Prisoner Walter Poller, quoted in Marco Pukrop, "Die SS-Karrieren von Dr. Wilhelm Berndt und Dr. Walter Döhrn. Ein Beitrag zu den unbekannten KZ-Ärzten der Vorkriegszeit," *Werkstatt Geschichte* 62 (2012), p. 79.

30 In his account of this episode (Gärtner and Kleinmann, *Doch der Hund*, p. 48), Fritz seems to imply that his "weeping and desperate" (*"weinender und verzweifelter"*) voice was an act.

31 Gustav's diary is hard to interpret here: *"(Am) nächsten Tag kriege (ich) einen Posten als Reiniger im Klosett, habe 4 Öfen zu heizen . . ."* The *Klosett* might have been the latrine in the little camp, or perhaps in the main camp barrack blocks, which had earlier been out of order due to a water shortage (Stein, *Buchenwald*, p. 86). The *Öfen* (ovens or furnaces) are harder to pinpoint; most likely they were part of the kitchens or the shower block. They were not crematorium ovens, which Buchenwald did not acquire until summer 1942 (ibid., p. 141).

Chapter 4: The Stone Crusher

1 Note of employment, undated, LJL; England and Wales census, 1911; description and details in passenger list, SS *Carinthia*, October 2, 1936, PNY; General Register Office 1939 Register, National Archives, Kew. Morris and Rebecca Brostoff were born in Białystok (now in Poland) around 1878 and emigrated to Britain prior to 1911. In 1939 they lived at 373 Street Lane.

2 Record card 46/01063-4, HOI. No record card for Richard Paltenhoffer from this time has been found, but he was presumably also put in Category C.

3 Wachsmann, *KL*, pp. 147–51; Cesarani, *Final Solution*, pp. 164–65; Wünschmann, *Before Auschwitz*, p. 186.

4 Arriving in Dachau on June 24, 1938, Richard Paltenhoffer was prisoner number 16865 (prisoner record, PGD). He was transferred to Buchenwald on September 23, 1938, where he was

assigned prisoner number 9520 and placed first in block 16, then block 14 (prisoner record, PGB).

5 Wachsmann, *KL*, pp. 181–84.

6 Ibid., p. 186.

7 A. R. Samuel, letter to David Makovski, May 25, 1939, LJW; marriage certificate, GRO; Montague Burton, letter to David Makovski, February 26, 1940, LJL; Nicholas Mark Burkitt, *British Society and the Jews* (2011), p. 108. The company was Rakusen Ltd., which still exists. Richard's first lodgings were at 9 Brunswick Terrace.

8 Biographical history, LJW; Anthony Grenville, "Anglo-Jewry and the Jewish Refugees from Nazism," *Association of Jewish Refugees Journal* (December 2012). The Leeds JRC was run by David Makovski, who ran a tailoring business in the city. He was known for a sometimes irascible temperament and a belief that each person should know their place in society and stick to it.

9 B. Neuwirth, letter to Richard Paltenhoffer, February 16, 1940; Control Committee, letter to Registrar of Marriages, February 20, 1940, LJL.

10 Gustav recorded all these imprecations in his poem "Quarry Kaleidoscope" (see later in this chapter).

11 Altogether, 1,235 prisoners died in Buchenwald in 1939, the majority of them in the last quarter of the year (Hackett, *Buchenwald Report*, p. 114).

12 The sequence of events at this period (including the precise assignments to barracks) differs somewhat between Gustav's diary and Fritz's recollections. The account given here reconciles the two.

13 The Goethe Oak was damaged by an Allied bomb in 1944 and was felled. However, its stump is still there.

14 Fritz Kleinmann, 1997 interview. Jewishness in itself was not sufficient cause to be sent to the camps until much later; at this time, the Nazi regime was focused on forcing Jews to emigrate, including those being held in the camps, who were released if they obtained the necessary emigration papers.

15 From "Quarry Kaleidoscope" by Gustav Kleinmann. I have translated Gustav's German as faithfully as possible:

> Klick-klack Hammerschlag,
> klick-klack Jammertag.
> Sklavenseelen, Elendsknochen,
> dalli und den Stein gebrochen.

16 Gustav's original:

> Klick-klack Hammerschlag,
> klick-klack Jammertag.
> Sieh nur diesen Jammerlappen
> winselnd um die Steine tappen.

17 Gustav and Fritz both record Herzog's first name as "Hans" but according to Stein (*Buchenwald*, p. 299) it was Johann. For other eyewitness accounts of Herzog's character and behavior see statements given in Hackett, *Buchenwald Report*, pp. 159, 174–5, 234. Although rumored to have later been murdered by a former prisoner, Herzog went on to have a long criminal career.

18 Gustav's original:

> Klatsch—er liegt auf allen Vieren,
> doch der Hund will nicht krepieren!

19 Gustav's original is more perfectly structured than my translation:

> Es rattert der Brecher tagaus und tagein,
> er rattert und rattert und bricht das Gestein,
> zermalt es zu Schotter und Stunde auf Stund'
> frißt Schaufel um Schaufel sein gieriger Mund.
> Und die, die ihn füttern mit Müh und mit Fleiß,
> sie wissen er frißt nur—doch satt wird er nie.
> Erst frißt er die Steine und dann frißt er sie.

Chapter 5: The Road to Life

1 Edith Kurzweil, *Nazi Laws and Jewish Lives* (2004), p. 153.

2 Report in Arad et al., *Documents*, pp. 143–44.

3 Rabinovici, *Eichmann's Jews*, pp. 87ff.

4 Passenger list, SS *Veendam*, January 24, 1940, PNY; United States census, 1940, NARA; Alfred Bienenwald, US passport application, 1919, NARA. Tini's cousins were Bettina Prifer and her brother, Alfred Bienenwald. Their mother, Netti, who was Hungarian-born, appears to have been a sister of Tini's mother, Eva née Schwarz (Bettina Bienenwald, birth record, October 20, 1899, Geburtsbuch and Geburtsanzeigen, IKA).

5 United States census, 1940, NARA.

6 US State Department memo, June 26 ,1940, in David S. Wyman, *America and the Holocaust* (1990), vol. 4, p. 1; also ibid., p. v.

7 Fritz and Gustav never understood where Tini got the money from, as she wasn't allowed to work. In fact she did get occasional jobs (letters to Kurt Kleinmann, 1941, DKK), and otherwise presumably depended on better-off relatives.

8 Gärtner and Kleinmann, *Doch der Hund*, p. 69; Buchenwald prisoner record card 1.1.5.3/6283376, ITS; Jeanette Rottenstein birth record, July 13, 1890, Geburtsbuch, IKA.

9 Fritz transferred into the garden detail on April 5, 1940 (prisoner record card 1.1.5.3/6283377, ITS).

10 Stein, *Buchenwald*, pp. 44–45, 307; Hackett, *Buchenwald Report*, p. 34. Hackmann's first name is variously given as Hermann and Heinrich. He was later convicted of embezzlement by the SS.

11 Gärtner and Kleinmann, *Doch der Hund*, pp. 47, 49. Fritz gives his height at this time as 145 cm (about 4 feet 9 inches). But in the 1938 family photograph, when he was fourteen, he is measurably only slightly shorter than the adult Edith (who was 5 feet 2 inches according to her passport: DPP). He must have grown a little in

the following eighteen months, so would have been over 5 feet tall (more than 152 cm) by late 1939.

12 Gustav Herzog was born in Vienna, January 12, 1908 (entry for Gustav Herzog, 68485, AMP).

13 Stefan Heymann was born in Mannheim, Germany, March 14, 1896 (entry for Stefan Heymann, 68488, AMP).

14 Anton Makarenko, *Road to Life: An Epic of Education (A Pedagogical Poem)*, vol. 2, ch. 1. Translation available online at www.marxists.org/reference/archive/makarenko/works/road2/ch01.html (retrieved February 4, 2020).

15 Fritz Kleinmann in Gärtner and Kleinmann, *Doch der Hund*, p. 54.

16 Hackett, *Buchenwald Report*, pp. 42, 336; Gärtner and Kleinmann, *Doch der Hund*, p. 55.

17 Stein, *Buchenwald* (German edition), p. 78.

18 Stein, *Buchenwald*, pp. 78–79.

19 Ibid., p. 90.

20 Gärtner and Kleinmann, *Doch der Hund*, p. 57. Schmidt's general temperament and habits are documented by many witnesses quoted in Hackett, *Buchenwald Report*.

Chapter 6: A Favorable Decision

1 Although they claimed to speak for "the people," in fact most Britons had had no notion of what a "fifth column" was until the *Daily Mail* began its campaign (Peter Gillman and Leni Gillman, *"Collar the Lot!" How Britain Interned and Expelled Its Wartime Refugees* (1980), pp. 78–79). The term "fifth column" originated during the Spanish Civil War (1936–39), when a general told the press that he had four columns of troops plus a "fifth column" within the enemy camp.

2 Roger Kershaw, "Collar the Lot! Britain's Policy of Internment During the Second World War," UK National Archives Blog

(2015). Most Jewish refugees were placed in Category C (exempt from internment) although 6,700 had been categorized B (subject to some restrictions), and 569 were judged a threat and interned. In fact there were spies and saboteurs at work in Britain, and dozens were eventually caught and convicted, but most were natural-born British citizens, not immigrants.

3 Gillman, *"Collar the Lot!,"* p. 153; Kershaw, "Collar the lot!"

4 Winston Churchill, House of Commons, June 4, 1940, *Hansard* vol. 364 c. 794.

5 Gillman, *"Collar the Lot!,"* pp. 167ff, 173ff; Kershaw, "Collar the Lot!"

6 Bernard Wasserstein, *Britain and the Jews of Europe, 1939–1945* (1999), p. 108.

7 Ibid., p. 83.

8 The address was 15 Reginald Terrace (various letters, LJL). At the time of their marriage, Richard had had lodgings at number 4 (marriage certificate, GRO). The Victorian houses in Reginald Terrace were demolished in the 1980s.

9 Leeds JRC, letter to Home Office, March 18, 1940, LJL. Mrs Green lived at 57 St. Martin's Garden.

10 Leeds and London JRC, letters, June 7 and 13, 1940, LJL.

11 Gillman, *"Collar the Lot!,"* pp. 113, 133. Edith had equipped herself with a certificate from her physician, Dr. Rummelsberg (April 24, 1940, LJL), presumably obtained for some purpose connected with her work or emigration application.

12 London, *Whitehall,* p. 171.

13 There is no record of where Richard Paltenhoffer was interned. His case file appears to have been among the majority which were later routinely destroyed by the Home Office (discovery.national-archives.gov.uk/details/r/C9246; retrieved February 4, 2020).

14 Joint Secretary, letter to Edith Paltenhoffer, August 30, 1940, LJL.

15 Joint Secretary, letter to Edith Paltenhoffer, September 4, 1940, LJL.

16 Home Office, letter to Leeds JRC, September 16, 1940, LJL.

17 Victor Cazalet, House of Commons, August 22, 1940, *Hansard* vol. 364 c. 1534.

18 Rhys Davies, House of Commons, August 22, 1940, *Hansard* vol. 364 c. 1529.

19 Home Office, letter to Leeds JRC, September 23, 1940, LJL. Richard's release had been approved on September 16 (record card 270/00271, HOI).

20 Quoted in Jerry Silverman, *The Undying Flame: Ballads and Songs of the Holocaust* (2002), p. 15.

21 Quoted in ibid., p. 15.

22 Manfred Langer in Hackett, *Buchenwald Report*, pp. 169–70.

23 Quoted in Silverman, *Undying Flame*, p. 15. Leopoldi survived the Holocaust, but Löhner-Beda was murdered in Auschwitz in 1942.

24 Hackett, *Buchenwald Report*, p. 42.

25 Fritz appears to have been transferred to the construction detail on August 20, 1940, after four months in the garden (prisoner record card 1.1.5.3/6283377, ITS).

26 In Gärtner and Kleinmann, *Doch der Hund*, p. 72.

27 The *Prominenten* of block 17 were of middling status. The Nazi regime kept its very highest-ranking political prisoners—former prime ministers, presidents, and monarchs of conquered countries—in isolation, often in special secret compounds within concentration camps. Buchenwald's was a walled compound in the spruce grove in front of the SS barracks.

28 Gedenkstätte Buchenwald, www.buchenwald.de/en/1218/ (retrieved February 2, 2020); Ulrich Weinzierl, *Die Welt*, April 1, 2005. Transferred to Dachau in October 1940, Fritz Grünbaum died there on January 14, 1941.

29 Tomas Plänkers, *Ernst Federn: Vertreibung und Rückkehr. Interviews zur Geschichte Ernst Federns und der Psychoanalyse* (1994), p. 158. Ernst Federn survived in Buchenwald until liberation in 1945; he continued his career in psychoanalysis and died in 2007.

30 In Gärtner and Kleinmann, *Doch der Hund*, p. 59.

31 Wachsmann, *KL*, pp. 224–25.

32 Ibid., p. 225. Cremation is forbidden in Jewish law, and cremated remains prohibited from cemeteries. However, exceptions are made for those cremated against their will, and ashes sent back from the concentration camps were permitted into Jewish cemeteries from the start.

33 Tini Kleinmann, letter to German Jewish Aid Committee, New York, March 1941, DKK.

34 Margaret E. Jones, letter to AFSC, November 1940, in Wyman, *America*, vol. 4, p. 3.

35 The consuls themselves, who didn't have to face the applicants, were generally callous, and even supported anti-Semitic immigration restrictions despite speaking publicly against Nazi anti-Semitism (Bat-Ami Zucker, *In Search of Refuge: Jews and US Consuls in Nazi Germany, 1933–1941* [2012], pp. 172–74). The Vienna consulate was more sympathetic than most, and willing to bend the rules a little (ibid., p. 167).

36 Tini Kleinmann, letter to German Jewish Aid Committee, New York, March 1941, DKK.

Chapter 7: The New World

1 This episode is based in part on interviews with Kurt Kleinmann, accounts written by him, and letters from Tini Kleinmann, July 1941, DKK; notes by Fritz Kleinmann, DRG; also data from passenger and crew list, SS *Siboney*, March 27, 1941, PNY.

2 There is an account of such a departure from Vienna in Ruth Maier, *Ruth Maier's Diary: A Young Girl's Life under Nazism*, trans. Jamie Bulloch (2009), pp. 112–13. If Kurt's train left in the evening, Tini and Herta would not have been allowed to accompany him to the station at all, due to the curfew; a non-Jewish friend or relative would have had to go with him.

3 Passenger and crew list, SS *Siboney*, March 27, 1941, PNY.

4 Efforts have been made by the author, by Kurt himself, and by the One Thousand Children organization to trace Karl Kohn and Irmgard Salomon, but no information has yet been found about their subsequent lives.

5 Description: records of the Selective Service System, Record Group Number 147: NARA.

Chapter 8: Unworthy of Life

1 In all the accounts of this murder (Gustav Kleinmann's diary; Emil Carlebach, Herbert Mindus in Hackett, *Buchenwald Report*, pp. 164, 171–72; Erich Fein and Karl Flanner, *Rot-Weiss-Rot in Buchenwald* [1987], p. 74) no mention is made of what triggered Abraham's actions.

2 Cesarani, *Final Solution*, p. 317; Stein, *Buchenwald*, pp. 81–83; Fritz Kleinmann in Gärtner and Kleinmann, *Doch der Hund*, pp. 77–79.

3 Herbert Mindus (in Hackett, *Buchenwald Report*, pp. 171–72) states that Hamber was in the construction detail and implies that the incident occurred on the SS garage site. However, that account was written four years later, whereas Gustav Kleinmann's diary is contemporary and probably more accurate, albeit less detailed; Gustav states that Hamber was in the haulage column (see also Fein and Flanner, *Rot-Weiss-Rot*, p. 74) and that the incident took place in an excavated part of the Economic Affairs department. Some accounts (Stein, *Buchenwald*, p. 288) date the incident to late 1940; in fact it was spring 1941.

4 Eduard's registered name appears to have been Edmund (Stein, *Buchenwald*, p. 298), but everyone knew him as Eduard (e.g. Fritz Kleinmann in Gärtner and Kleinmann, *Doch der Hund*, p. 81; Herbert Mindus in Hackett, *Buchenwald Report*, p. 171).

5 Reported by Emil Carlebach in Hackett, *Buchenwald Report*, p. 164.

6 Ibid.

7 Stein, *Buchenwald*, p. 298.

8 Gustav is enigmatic here; he uses the word *"Aktion,"* meaning a "campaign" or "special operation," implying that he had in mind some kind of concerted resistance among the haulage column, led by Eduard Hamber. However, his writing is extremely elliptical—probably because, while keeping a diary would no doubt be fatal for him if found out, the consequences would be even worse if it contained evidence of anti-SS activities.

9 Tini Kleinmann, letter to Kurt Kleinmann, July 15, 1941, DKK.

10 Order of May 14, 1941, quoted in Gold, *Geschichte der Juden*, pp. 106–7.

11 Cesarani, *Final Solution*, p. 443.

12 Rabinovici, *Eichmann's Jews*, p. 136.

13 Cesarani, *Final Solution*, p. 418.

14 Tini Kleinmann, letter to Kurt Kleinmann, August 5, 1941, DKK.

15 Tini Kleinmann, letter to Kurt Kleinmann, July 15, 1941, DKK.

16 Tini Kleinmann, letters to Kurt Kleinmann, July–August 1941, DKK.

17 Prisoner record cards 1.1.5.3/6283389, 1.1.5.3/6283376, ITS. The record indicates four packages signed for during 1941—one each for Gustav and Fritz on May 3, one for Fritz on October 22, and one for Gustav on November 16. All contained items of clothing.

18 Gustav writes: *"Wir sind die Unzertrennlichen"*—"We are the inseparables." There is no exact equivalent of *Unzertrennlichen* in English. In German it is used for the bird species known in English as lovebirds, and is also the German title of the David Cronenberg film *Dead Ringers*.

19 William L. Shirer, quoted in Cesarani, *Final Solution*, p. 285.

20 Stein, *Buchenwald*, pp. 124–26; Wachsmann, *KL*, pp. 248–58; Cesarani, *Final Solution*, pp. 284–86.

21 SS-Doctor Waldemar Hoven, quoted in Stein, *Buchenwald*, p. 124.

22 Gustav gives the date as August 1941; he is normally totally reliable on dates, but it seems that he described the events of spring

and summer 1941 retrospectively—probably at the end of the year—and his chronology and figures are sometimes unreliable for this period.

23 Stein, *Buchenwald*, p. 59.

24 Otto Kipp in Hackett, *Buchenwald Report*, p. 212.

25 SS nurse Ferdinand Römhild, quoted in Stein, *Buchenwald*, p. 126.

26 It was true that many of the leading Bolshevik revolutionaries of 1917 had been Jews, and it was also true that the Soviet regime had liberated Russian Jews from the anti-Semitic repression of the tsars. But the alleged connection between Jewishness and communism was just a fantasy in the minds of Nazi ideologues, a banal modern equivalent of the blood libel.

27 Wachsmann, *KL*, p. 260.

28 Gustav Kleinmann's diary says that this occurred on June 15. This is impossible, as war between Germany and the USSR did not begin until June 22. This is another instance of his misattributing the date of a 1941 event due to writing about it from memory (see note 22 above). Aside from the date, all the other details of his account are corroborated by multiple sources.

29 Stein, *Buchenwald*, pp. 121–24; Hackett, *Buchenwald Report*, pp. 236ff; Wachsmann, *KL*, pp. 258ff. "Commando 99" was a reference to the telephone number of the stables.

30 Stein, *Buchenwald*, p. 85; Wachsmann, *KL*, pp. 277ff.

31 Stein, *Buchenwald*, pp. 121–23.

32 Gustav uses the word "*Justifizierungen*," a euphemism sometimes used for judicial murder, for which there is no exact English equivalent—adjustment, judgment, or adjudication are near translations.

33 Fritz Kleinmann in Gärtner and Kleinmann, *Doch der Hund*, p. 21n.

34 Wachsmann, *KL*, pp. 270–71. A similar effect had been observed among the Einsatzgruppen death squads on the eastern front; shooting large numbers of victims at close range over a long period traumatized even hardened, dedicated SS men (Cesarani,

Final Solution, p. 390). This was one of the reasons for the move toward using gas chambers in concentration camps, and forcing teams of prisoners—the Sonderkommandos—to handle the victims.

35 Stein, *Buchenwald*, pp. 58–59; witness statements in Hackett, *Buchenwald Report*, pp. 71, 210, 230; Wachsmann, *KL*, p. 435.

36 Stein, *Buchenwald*, p. 58.

37 Ibid., pp. 200–203; Wachsmann, *KL*, p. 435. The typhus serums with which they were injected were being developed jointly by the SS, the IG Farben chemical corporation, and the Wehrmacht, with the aim of producing a vaccine for German troops serving in Eastern Europe, where typhus was endemic.

38 Fritz Kleinmann in Gärtner and Kleinmann, *Doch der Hund*, pp. 79–80.

39 *Völkischer Beobachter*, quoted in Cesarani, *Final Solution*, p. 421.

40 Rees, *Holocaust*, p. 231; Cesarani, *Final Solution*, pp. 421ff; notes on accession no. 2005.506.3, United States Holocaust Memorial Museum, collections.ushmm.org/search/catalog/irn523 540 (retrieved February 20, 2020).

41 Rabinovici, *Eichmann's Jews*, pp. 110–11.

42 Tini Kleinmann, letter to Samuel Barnet, July 19, 1941, DKK.

43 Fritz Kleinmann in Gärtner and Kleinmann, *Doch der Hund*, p. 83.

44 Rees, *Holocaust*, p. 231; Cesarani, *Final Solution*, pp. 422ff.

45 Order from Heinrich Müller, RSHA, October 23, 1941, in Arad et al., *Documents*, pp. 153–4.

Chapter 9: A Thousand Kisses

1 Michael Dror, "News from the Archives," *Yad Vashem Jerusalem* 81 (2016), p. 22. Arnold Frankfurter died in Buchenwald on either February 14 (Felix Czeike, *Historisches Lexikon Wien* (1992–97), vol. 2, p. 357) or March 10/19, 1942 (Felicitas Heimann-Jelinek,

Lothar Höbling, and Ingo Zechner, *Ordnung muss sein: Das Archiv der Israelitischen Kultusgemeinde Wien* [2007], p. 152). He married Gustav Kleinmann and Tini Rottenstein in Vienna on May 8, 1917 (Dieter J. Hecht, "'Der König rief, und alle, alle kamen': Jewish Military Chaplains on Duty in the Austro-Hungarian Army During World War I," *Jewish Culture and History* 17/3 [2016]: pp. 209–10).

2 Fritz Kleinmann in Gärtner and Kleinmann, *Doch der Hund*, p. 82.

3 Cesarani, *Final Solution*, pp. 445–49.

4 Stein, *Buchenwald*, p. 128.

5 Hermann Einziger in Hackett, *Buchenwald Report*, p. 189.

6 Gustav is specific that Greuel was the SS sergeant involved. Confusingly, he seems to say that this incident occurred on a "gravel transport from the crusher." However, it occurs within the context of his writing about transporting tree trunks from the forest. Presumably his team were doing both jobs alternately. The fact that some of Gustav's men were not carrying anything on this occasion suggests that this occurred during log carrying rather than gravel transport (which would have been by wagon).

7 Robert Siewert and Josef Schappe in Hackett, *Buchenwald Report*, pp. 153, 160.

8 Fritz says that Leopold Moses went to Natzweiler in 1941 (Gärtner and Kleinmann, *Doch der Hund*, p. 50). However, at that time Natzweiler had only a very small number of prisoners (transferred from Sachsenhausen); it began to receive large transports in spring 1942 (Jean-Marc Dreyfus in Geoffrey P. Megargee, ed., *The United States Holocaust Memorial Museum Encyclopedia of Camps and Ghettos, 1933–1945* [2009], vol. 1B, p. 1,007).

9 Fritz Kleinmann in Gärtner and Kleinmann, *Doch der Hund*, p. 82. (Tini's original letter, which Fritz never saw, was not preserved.)

10 Former Soviet territory under German rule was divided into Reichskommissariat Ostland and Reichskommissariat Ukraine.

Beyond these regions was a still larger war zone at the rear of the German front line.

11 The instructions for deportees from the Altreich and Ostmark to the Ostland are outlined in Cesarani, *Final Solution*, p. 428; Christopher Browning, *The Origins of the Final Solution* (2005), p. 381; and a memorandum in Arad et al., *Documents*, pp. 159–61. The instruction leaflet issued to transport supervisors in Vienna is quoted in full in Gold, *Geschichte*, pp. 108–9. The narrative of a deportee is given in the testimony of Viennese survivor Wolf Seiler (deported May 6, 1942), document 854, DOW.

12 Transports of Jews to the Ostland began in November 1941; there were seven that month from various German cities, including one from Vienna (Alfred Gottwaldt, "Logik und Logistik von 1300 Eisenbahnkilometern" in Waltraud Barton, ed., *Ermordet in Maly Trostinec: die österreichischen Opfer der Shoa in Weißrussland* [2012], p. 54). The program was interrupted due to the logistical demands of the Wehrmacht and resumed in May 1942; between then and October there were nine transports from Vienna, leaving weekly in late May and June (ibid.; see also Alfred Gottwaldt and Diana Schulle, *Die «Judendeportationen» aus dem Deutschen Reich 1941–1945* [2005], pp. 230ff; Irene Sagel-Grande, H. H. Fuchs and C. F. Rüter, *Justiz und NS-Verbrechen: Sammlung Deutscher Strafurteile wegen Nationalsozialistischer Tötungsverbrechen 1945–1966: Band XIX* [1978], pp. 192–96).

13 Gärtner and Kleinmann, *Doch der Hund*, p. 69; Buchenwald prisoner record card 1.1.5.3/6283376, ITS.

14 Bertha's husband had been killed in the First World War, and she never remarried (Bertha Rottenstein birth record, April 29, 1887, Geburtsbuch, IKA; *Lehmann's Adressbuch* for Vienna for 1938, WLO; casualty reports, *Illustrierte Kronen Zeitung*, June 4, 1915, p. 6; K.u.k. Kriegsministerium, *Verlustliste Nr 209 ausgegeben am 13.7.1915* [1915], p. 54).

15 How long Tini and Herta Kleinmann were held in the *Sammellager* (holding camp) isn't known; some deportees waited a week

or more, but as Tini's and Herta's deportation serial numbers were quite high (see note below), they were presumably notified quite late and would not have been held for long.

16 Loading could take over five hours (e.g., police report on transport Da 230, October 1942, DOW).

17 The deportees are listed in the Gestapo departure list for Transport 26 (Da 206), June 9, 1942, 1.2.1.1/11203406, ITS; limited data also available in Erfassung der Österreichischen Holocaustopfer (Database of Austrian victims of the Holocaust), DOW and YVS.

18 Tini Rottenstein was born January 2, 1893, in the apartment building at Kleine Stadtgutgasse 6, near the Praterstern (Geburtsbuch 1893, IKA).

19 The Aspangbahnhof was demolished in 1976. A small square— Platz der Opfer der Deportation (Deportation Victims' Square)—now stands on the site, along with a memorial to the thousands of deportees who left Vienna from the station.

20 The route is given in Alfred Gottwaldt, "Logik und Logistik von 1300 Eisenbahnkilometern" in Barton, *Ermordet*, pp. 48–51. Timings are estimated from the Vienna police report on transport Da 230, October 1942, DOW.

21 When the war began, the SS-Totenkopfverbände (Death's Head units) division was placed under the overall command of the Waffen-SS. Veteran guard personnel were sent to fight on the Eastern Front. They were replaced in the camps by new volunteers and conscripts. The Death's Head insignia was worn on the caps of all SS men, but only the SS-TV wore it on their collar tabs also.

22 Sipo-SD was the informal name of the combined units of the SS Sicherheitspolizei (Sipo, security police) and Sicherheitsdienst (SD, intelligence). The Sipo, which combined the Gestapo and the criminal police, was defunct by this time, having been absorbed into the Reich Main Security Office (RSHA), but the term was still used for the combined police-SD units operating in the eastern territories.

23 Testimony of survivor Wolf Seiler (deported May 6, 1942), document 854, DOW; testimony of Isaak Grünberg (deported October 5, 1942), quoted in Alfred Gottwaldt, "Logik und Logistik von 1300 Eisenbahnkilometern" in Barton, *Ermordet*, p. 49.

24 Alfred Gottwaldt, "Logik und Logistik von 1300 Eisenbahnkilometern" in Barton, *Ermordet*, p. 51.

25 The transport that left Vienna on Tuesday, June 9, is recorded as arriving at Minsk on either Saturday, June 13 or Monday, June 15; rail records indicate the former date, whereas a report by SS-Lieutenant Arlt (June 16, 1942: file 136 M.38, YVP) indicates the latter. Holocaust deniers have taken this discrepancy as casting doubt on the massacres at Maly Trostinets. In fact it was due to industrial relations; as of May 1942, railway workers in Minsk were not required to work weekends, and trains arriving on a Saturday were parked at Kojdanów station outside the city until Monday morning (Alfred Gottwaldt, "Logik und Logistik von 1300 Eisenbahnkilometern" in Barton, *Ermordet*, p. 51).

26 Tini Kleinmann, letter to Kurt Kleinmann, August 5, 1941, DKK.

27 Sources used here include secondary accounts (Sybille Steinbacher, "Deportiert von Wien nach Minsk" in Barton, *Ermordet*, pp. 31–38; Sagel-Grande et al., *Justiz*, pp. 192–96; Christian Gerlach, *Kalkulierte Morde: Die deutsche Wirtschafts- und Vernichtungspolitik in Weißrußland 1941 bis 1944* [1999], pp. 747–60; Petra Rentrop, "Maly Trostinez als Tatort der «Endlösung»" in Barton, *Ermordet*, pp. 57–71; Mark Aarons, *War Criminals Welcome: Australia, a Sanctuary for Fugitive War Criminals Since 1945* [2001], pp. 71–6), official reports (SS-Lieutenant Arlt, June 16, 1942: file 136 M.38, YVP), and personal testimonies of survivors (Wolf Seiler, document 854, DOW; Isaak Grünberg, quoted in various preceding citations).

28 Petra Rentrop, "Maly Trostinez als Tatort der «Endlösung»" in Barton, *Ermordet*, p. 64.

29 Cesarani, *Final Solution*, pp. 356ff.

30 Sybille Steinbacher, "Deportiert von Wien nach Minsk" in Barton, *Ermordet*, pp. 31–38; Sagel-Grande et al., *Justiz*, pp. 192–96; Gerlach, *Kalkulierte Morde*, pp. 747–60; Petra Rentrop, "Maly Trostinez als Tatort der «Endlösung»" in Barton, *Ermordet*, pp. 57–71. Maly Trostinets concentration camp is rarely mentioned in general Holocaust histories; even the mammoth four-volume *United States Holocaust Memorial Museum Encyclopedia of Camps and Ghettos* (ed. Megargee) does not have an entry for it, just a few references in the entry for the Minsk ghetto (vol. 2B, pp. 1,234, 1,236). There are many variant spellings of the name in the literature—in modern Belarusian it is Maly Trościeniec; other variants include Trostenets, Trostinets, Trostinec, Trostenez, Trastsianiets, Trascianec. In German it is sometimes referred to as Klein Trostenez.

31 Testimony of Wolf Seiler, document 854, DOW.

32 Sagel-Grande et al., *Justiz*, p. 194.

33 Aarons, *War Criminals*, pp. 72–74.

34 Sagel-Grande et al., *Justiz*, p. 194.

35 Aarons, *War Criminals*, pp. 72–74.

36 Petra Rentrop, "Maly Trostinez als Tatort der «Endlösung»" in Barton, *Ermordet*, p. 65. There may in fact have been up to eight gas vans in Belarus, but only three or four appear to have been used at Maly Trostinets (Gerlach, *Kalkulierte Morde*, pp. 765–66).

37 Sagel-Grande et al., *Justiz*, pp. 194–95.

38 SS-Lieutenant Arlt, June 16, 1942: file 136 M.38, YVP.

39 Tini refers to this rowing outing and to her own childhood in her last letter to Kurt, July 15,1941, DKK.

40 Altogether, according to the Dokumentationsarchiv des österreichischen Widerstandes (www.doew.at), about 9,000 Jews were deported from Vienna to Maly Trostinets. Only seventeen are known to have survived. The total numbers killed at Maly Trostinets are not known for certain, but it is estimated that over 200,000 German, Austrian, and Belarusian Jews and Soviet prisoners of war were murdered there between 1941 and 1943, when

the camp was shut down (Martin Gilbert, *The Holocaust: The Jewish Tragedy* [1986], p. 886 n. 38).

Chapter 10: A Journey to Death

1 An account of this incident given after the war by prisoner Hermann Einziger (in Hackett, *Buchenwald Report*, p. 189) states that it occurred in April, and that the labor detail were carrying the logs to the camp by hand. However, Gustav's diary (which returns to its usual chronological reliability in 1942) suggests that it was later in the year (mid to late summer) and that the logs were being loaded onto a wagon. Einziger says Friedmann was from Mannheim; Gustav says he was from Kassel. Neither offers any further details about him.

2 The ban on Jews being admitted to the infirmary had been lifted at some point; the precise date isn't known.

3 Stein, *Buchenwald*, pp. 138–39; Ludwig Scheinbrunn in Hackett, *Buchenwald Report*, pp. 215–16.

4 Stein, *Buchenwald*, pp. 36–37; Hackett, *Buchenwald Report*, p. 313.

5 Order of October 5, 1942, quoted in Stein, *Buchenwald*, p. 128.

6 Stein, *Buchenwald*, pp. 128–29.

7 This was a full week after the drafting of the list on October 8 (Stefan Heymann in Hackett, *Buchenwald Report*, p. 342).

8 Fritz doesn't explain this in his memoir. The SS would not require such an affirmation, so it was perhaps for Siewert's benefit, in case he was accused of being complicit or of forcing Fritz to go.

9 Fritz (in Gärtner and Kleinmann, *Doch der Hund*, p. 86) says there were eighty men to a wagon; however, Commandant Pister had ordered a train from the railway company consisting of ten cattle/freight wagons and one passenger carriage for SS personnel (Stein, *Buchenwald Report*, pp. 128–29). Fritz gives the date of departure as October 18, and of arrival at Auschwitz as October 20, off by one day.

10 Gustav uses the stock expression "*Himmelfahrtskommando*," which translates literally as "trip to Heaven mission" and is the German equivalent of "suicide mission" or "kamikaze order."

Chapter 11: A Town Called Oświęcm

1 Men in Austria–Hungary were drafted into the army in the spring of the year in which they turned twenty-one, serving three years followed by ten years in the reserves (James Lucas, *Fighting Troops of the Austro-Hungarian Army, 1868–1914* [1987], p. 22). Gustav Kleinmann turned twenty-one on May 2, 1912. The kaiserlich und königlich (k.u.k.) Armee (Imperial and Royal Army) was made up of troops from all over the empire.

2 Lucas, *Fighting Troops*, pp. 25–26.

3 The 12th Infantry Division was part of First Army's support force, and was attached to X Corps for the advance.

4 John R. Schindler, *Fall of the Double Eagle: The Battle for Galicia and the Demise of Austria–Hungary* (2015), p. 171. In 1914, the north and west of what is now Poland was part of the German Empire, and the south (comprising Galicia) belonged to Austria-Hungary. Central modern Poland (including Warsaw) was part of the Russian Empire. Thus Austria's border with Russia was to the north and east.

5 Ibid., pp. 172ff.

6 Ibid., pp. 200–239.

7 Alexander Watson, *Ring of Steel: Germany and Austria–Hungary at War, 1914–1918* (2014), pp. 193–95.

8 Ibid., pp. 200–201; Andrew Zalewski, *Galician Portraits: In Search of Jewish Roots* (2012), pp. 205–6.

9 John Keegan, *The First World War* (1998), p. 192.

10 Gemeinesames Zentralnachweisbureau, *Nachrichten über Verwundete und Kranke Nr 190 ausgegeben am 6.1.1915* (1915), p. 24; *Nr 203 ausgegeben am 11.1.1915* (1915), p. 25. The exact circumstances of Gustav's

wound are not known, other than that he was shot. The two reports cited indicate respectively that he was shot in the left lower leg ("*linken Unterschenkel*," January 6, Biala) and left forearm ("*linken Unterarm*," January 11, Oświęcim). Simultaneous wounds in the left arm and left leg sometimes happened when a soldier was kneeling to fire his rifle. Such wounds would probably have occurred during an attack or raid rather than in trenches.

11 Robert Jan van Pelt and Debórah Dwork, *Auschwitz: 1270 to the Present* (1996), p. 59.

12 Ibid.

13 The report describing Gustav's actions (Award application, 3 Feldkompanie, Infanterieregiment 56, February 27, 1915, BWM) indicates that this was entirely on Gustav's and Aleksiak's own initiative, which suggests that their sergeant and/or platoon officer was absent, most likely killed in the assault.

14 Austro-Hungarian Army report, February 26, 1915, *Amtliche Kriegs-Depeschen*, vol. 2 (Berlin: Nationaler Verlag, 1915): reproduced online at stahlgewitter.com/15_02_26.htm (retrieved February 4, 2020).

15 Award application, 3 Feldkompanie, Infanterieregiment 56, February 27, 1915, BWM.

16 *Wiener Zeitung*, April 7, 1915, pp. 5–6. Altogether, nineteen men of the 56th were awarded the Silver Medal for Bravery 1st Class (Silberne Tapferkeitsmedaille erster Klasse) while ninety-seven received the 2nd Class.

17 K.u.k. Kriegsministerium, *Verlustliste Nr 244 ausgegeben am 21.8.1915* (1915), p. 21. The official list doesn't specify how Gustav received this wound or where it was located (nor indeed which hospital he was in); he is merely listed as "*verwundeten.*" Family oral history says it was in the lung.

18 This is the substance of speeches given by Rabbi Arnold Frankfurter at this time, including at weddings, as quoted by Hecht in "Der König rief," pp. 212–13, which specifically mentions Gustav and Tini's wedding.

19 Watson, *Ring of Steel*, pp. 503–6.

20 Grünberg was registered as a bricklayer's apprentice on the Auschwitz intake (arrivals list, October 19, 1942, ABM).

21 Prior to 1944, when a rail spur and loading ramp were constructed in the Birkenau camp, prisoners arriving at Auschwitz disembarked at a spur near Auschwitz I, and prior to that at the station in the town, and marched to the camps.

22 Danuta Czech, *Auschwitz Chronicle: 1939–1945* (1990), p. 255.

23 There were 405 men on the transport list, but only 404 were admitted to Auschwitz (ibid., p. 255). Presumably one had died en route.

24 Auschwitz I later acquired a dedicated admissions building outside the camp entrance (van Pelt and Dwork, *Auschwitz*, pp. 222–25; Czech, *Auschwitz Chronicle*, p. 601). Prior to that there were only the regular facilities inside the camp.

25 The first gassings in Germany, using trucks and gas chambers, had occurred in 1939, as part of the T4 euthanasia program (Cesarani, *Final Solution*, pp. 283–85). The first experimental gassings with Zyklon B at Auschwitz were done in August 1941 in Auschwitz I; large, specialized gas chambers/crematoria came into use in Auschwitz-Birkenau in early 1942 (Wachsmann, *KL*, pp. 267–68, 301–2; Franciszek Piper in Megargee, *USHMM Encyclopedia*, vol. 1A, pp. 206, 210). By late 1942, rumors about gassings had spread through the concentration camp system and among local populations.

26 "*Eine Laus dein Tod*"—this message was painted on walls throughout the Auschwitz complex.

27 Delousing of uniforms was done by fumigation with Zyklon B. This was the original intended purpose of this poison gas, which the SS adapted for use in the killing gas chambers. For that purpose, they asked the manufacturer (a subsidiary of IG Farben) to remove the noxious warning smell which was normally added to it (Peter Hayes, *Industry and Ideology: IG Farben in the Nazi Era* [2001], p. 363).

28 The first recipients of tattooed numbers were Soviet POWs, beginning in autumn 1941. The SS experimented with a stamping device early on, but it hadn't worked very well (Wachsmann, *KL*, p. 284). No other concentration camp used tattooing.

29 Arrivals list, October 19, 1942, ABM.

30 The numbering Auschwitz I, II, and III was not introduced until November 1943 (Florian Schmaltz in Megargee, *USHMM Encyclopedia*, vol. 1A, p. 216), but is used here for clarity and consistency.

31 Franciszek Piper in Megargee, *USHMM Encyclopedia*, vol. 1A, p. 210. Auschwitz-Birkenau (Auschwitz II) began construction in October 1941 and was operational in early 1942.

32 Gustav uses the phrase *"schwarze Mauer"* rather than the more commonly used *"schwarze Wand."* Both mean the same. Its name came from the black-painted screen that protected the brick wall from bullet strikes.

33 Czech, *Auschwitz Chronicle*, p. 259.

34 Höss, quoted in Hermann Langbein, *People in Auschwitz*, trans. Harry Zohn (2004), pp. 391–92.

35 Quoted in Langbein, *People*, p. 392. At around this time, SS-Sergeant Gerhard Palitzsch became increasingly unbalanced, due to the death of his wife. They lived in a house near the camp, and Palitzsch, who was involved in corruption, obtained clothes stolen from the prisoners in Birkenau. In October 1942 his wife contracted typhus—probably from lice carried in these clothes—and died. Palitzsch took to drinking heavily and his behavior became erratic (ibid., pp. 408–10).

36 Czech, *Auschwitz Chronicle*, pp. 255–60.

37 Ibid., p. 261. The 186 women from Ravensbrück were declared fit and assigned work separately from the men (ibid., pp. 261–62).

38 In Gärtner and Kleinmann, *Doch der Hund*, p. 90. Fritz says that they stayed only a week in Auschwitz I, and in their testimony to the Frankfurt trials both he and Gustav stated the time as eight days (Abt 461 Nr 37638/84/15904–6; Abt 461 Nr 37638/83/15661–3,

FTD); in fact it was eleven days (Czech, *Auschwitz Chronicle*, pp. 255, 260–61).

39 The truth of this is uncertain. There was a heavy demand for workers for construction of the new Monowitz camp, and the records imply that the intention all along had been to send the transferred prisoners to work there (Czech, *Auschwitz Chronicle*, p. 255). However, the record is unclear, and Fritz and Gustav had the impression that they were all slated for execution. Certainly that was the purpose of their selection at Buchenwald—hence the retention of construction workers.

Chapter 12: Auschwitz-Monowitz

1 At this time the camp was officially referred to as the Buna labor camp (or as "Camp IV" by IG Farben management—see Bernd C. Wagner, *IG Auschwitz: Zwangsarbeit und Vernichtung von Häftlingen des Lagers Monowitz 1941–1945* [2000], p. 96). Later it became known as Monowitz concentration camp or Auschwitz III. The later names are used here for clarity and consistency.

2 By early September 1942 the Monowitz camp had been completely laid out, but construction hadn't progressed beyond a small number of barracks (between two and eight, according to sources). The rest of the camp buildings had been delayed in order to expedite construction of the Buna Werke factory. The camp officially opened for reception of prisoners on October 28 (ibid., pp. 95–97).

3 The Buna Werke was named after buna, the synthetic rubber intended to be produced there; among other applications, rubber was vital in aircraft and vehicle manufacture—e.g., tires and various shock-absorbing components.

4 Florian Schmaltz in Megargee, *USHMM Encyclopedia*, vol. 1A, pp. 216–17; Gärtner and Kleinmann, *Doch der Hund*, p. 92. Eventually camp inmates would make up about a third of the Buna

Werke's total workforce, the rest made up of paid workers from Germany or occupied countries (Hayes, *Industry*, p. 358), many of whom would be drafted labor from enforced schemes such as France's Service du Travail Obligatoire.

Chapter 13: The End of Gustav Kleinmann, Jew

1 Wachsmann, *KL*, pp. 49–52; Joseph Robert White in Megargee, *USHMM Encyclopedia*, vol. 1A, pp. 64–6. Esterwegen and the other Emsland camps were shut down in 1936.

2 Lehmann directory name listings, 1891, WLO; Alice Teichova, "Banking in Austria" in Manfred Pohl, ed., *Handbook on the History of European Banks* (1994), p. 4.

3 Wagner, *IG Auschwitz*, p. 107.

4 The term was used in other camps as well. Its origin is not known. (See Yisrael Gutman in Yisrael Gutman and Michael Berenbaum, eds., *Anatomy of the Auschwitz Death Camp* (1994), p. 20; Wachsmann, *KL*, pp. 209–10, 685 n. 117; Wladyslaw Fejkiel quoted in Langbein, *People*, p. 91.) By the time the concentration camps were liberated in 1944–45, most long-term prisoners had been turned into *Muselmänner*, and they became emblematic of the Holocaust's victims. But they existed in the camps as early as 1939.

5 Hayes, *Industry*, p. 358.

6 Herzog was a clerk from mid-1943, and head of the office from January to October 1944 (Herzog, Frankfurt trials statement, Abt 461 Nr 37638/84/15891–2, FTD).

7 Detailed plan and layout of buildings by Irena Strzelecka and Piotr Setkiewicz, "Bau, Ausbau und Entwicklung des KL Auschwitz" in Wacław Długoborski and Franciszek Piper, eds., *Auschwitz 1940–1945: Studien der Geschichte des Konzentrations- und Vernichtungslagers Auschwitz* (1999), vol. 1, pp. 128–30.

8 Wachsmann, *KL*, p. 210.

9 Primo Levi, who was a prisoner in Monowitz from February 1944, said of block 7 that "no ordinary Häftling [prisoner] has ever entered" (Primo Levi, *Survival in Auschwitz and The Reawakening: Two Memoirs* [1986], p. 32).

10 Wagner, *IG Auschwitz*, pp. 117, 121–22; Langbein, *People*, pp. 150–51.

11 Quoted in Wachsmann, *KL*, p. 515.

12 Wagner, *IG Auschwitz*, pp. 121–22.

13 Ibid., p. 117.

14 Freddi Diamant, quoted in Langbein, *People*, p. 151.

15 Irena Strzelecka and Piotr Setkiewicz, "Bau, Ausbau und Entwicklung des KL Auschwitz" in Długoborski and Piper, *Auschwitz 1940–1945*, vol. 1, p. 135.

16 By the end of 1943 Auschwitz had three satellite camps dedicated to coal mining: Fürstengrube, Janinagrube, and Jawischowitz. They ranged from around 15 to 100 km, distant from the main Auschwitz camp (entries in Megargee, *USHMM Encyclopedia*, vol. 1A, pp. 221, 239, 253, 255).

17 Wagner, *IG Auschwitz*, p. 118. Always a cunning operator and well liked by the SS, within a few weeks Windeck wangled himself a position as camp senior in the Birkenau men's camp.

Chapter 14: Resistance and Collaboration: The Death of Fritz Kleinmann

1 Wachsmann, *KL*, pp. 206–7.

2 Fritz Kleinmann in Gärtner and Kleinmann, *Doch der Hund*, p. 108.

3 The following details are described at length by Fritz Kleinmann in ibid., pp. 108–12.

4 Langbein, *People*, p. 142; Irena Strzelecka and Piotr Setkiewicz, "Bau, Ausbau und Entwicklung des KL Auschwitz" in Długoborski and Piper, *Auschwitz 1940–1945*, vol. 1, p. 128.

5 Hayes, *Industry*, pp. 361–62.

6 Fritz Kleinmann in Gärtner and Kleinmann, *Doch der Hund*, p. 112; author's translation.

7 Florian Schmaltz in Megargee, *USHMM Encyclopedia*, vol. 1A, p. 217.

8 Henryk Swiebocki, "Die Entstehung und die Entwicklung der Konspiration im Lager" in Długoborski and Piper, *Auschwitz 1940–1945*, vol. 4, pp. 150–53.

9 Pierre Goltman, *Six Mois en Enfer* (2011), pp. 89–90.

10 Fritz states that he worked as a *"Transportarbeiter,"* "transport worker" (Gärtner and Kleinmann, *Doch der Hund*, p. 113), but doesn't elucidate; this was quite a broad label, and probably denotes fetching and carrying for locksmith technicians within the factory.

11 Hermann Langbein in Gutman and Berenbaum, *Anatomy*, pp. 490–91; Henryk Swiebocki, "Die Entstehung und die Entwicklung der Konspiration im Lager" in Długoborski and Piper, *Auschwitz 1940–1945*, vol. 4, pp. 153–54.

12 Florian Schmaltz in Megargee, *USHMM Encyclopedia*, vol. 1A, p. 217.

13 Langbein, *People*, p. 329.

14 Ibid., pp. 31, 185, 322, 329–35.

15 In his memoir and interview, Fritz says only that he was taken to the Political Department, without specifying whether it was the main department at Auschwitz I or the sub department in Monowitz. The involvement of Grabner and the seriousness of the charge suggest that it was probably the main department. On the other hand, at the end of the interrogation he says that Grabner "went back to Auschwitz with the civilian" (Gärtner and Kleinmann, *Doch der Hund*, p. 114); but he also writes that Taute and Hofer took him "back to the camp" (ibid.) which again suggests Auschwitz I as the scene of the torture. Overall, the balance of evidence favors the latter. In his 1963 statement for the Frankfurt trials (Abt 461 Nr 37638/83/15663, FTD), Fritz stated that this incident occurred in June 1944; as Grabner left Auschwitz in late 1943, this is probably a transcription error for June 1943.

16 Wagner, *IG Auschwitz*, pp. 163–92; Irena Strzelecka and Piotr Setkiewicz, "Bau, Ausbau und Entwicklung des KL Auschwitz" in Długoborski and Piper, *Auschwitz 1940–1945*, vol. 1, p. 128.

17 The entry recording Fritz Kleinmann's death has not come to light; presumably it was among the majority of Auschwitz records destroyed before the liberation of the camp. Some hospital registers have survived (and have the format described), but this one is apparently lost.

18 In his published recollections Fritz makes no mention of his suicidal thoughts at this time, but in his 1997 interview he describes them at some length and with strong emotion.

19 Fritz is unclear about exactly how long it was before his father was told about his survival. In his written memoir, he implies that it was shortly after his transfer from the hospital to block 48, whereas in his 1997 interview he is vague, implying that through necessity the secret was kept for a long time.

20 Czech, *Auschwitz Chronicle*, pp. 537, 542.

21 Langbein, *People*, p. 40; Wachsmann, *KL*, pp. 388–89; Czech, *Auschwitz Chronicle*, pp. 537, 812.

22 Prisoner resistance report, December 9, 1943, quoted in Czech, *Auschwitz Chronicle*, p. 542.

Chapter 15: The Kindness of Strangers

1 The version of this incident given by Fritz Kleinmann differs in some details from the version in Gustav's diary, and both differ from the Gestapo records (as quoted in Czech, *Auschwitz Chronicle*, pp. 481–82). The account given here is a synthesis of the three.

2 Gustav recorded in his diary that both Eisler and Windmüller were shot (cf. Czech, *Auschwitz Chronicle*, p. 482); presumably this was the story that came back to Monowitz at the time.

3 Not to be confused with the Rote Hilfe eV, a socialist aid organization founded in 1975. The original Rote Hilfe was founded in 1921 as an affiliate of the International Red Aid. It was banned under the Nazis, and later disbanded. Many of its activists ended up in concentration camps.

4 It is not known exactly what Alfred Wocher's duties were on the eastern front, or what unit he was in, but it is hard to believe that he was unaware of the mass murders of Jews carried out there. By no means were the Waffen-SS and Einsatzgruppen the only organizations involved; Wehrmacht units took part too, and even if Wocher was nowhere near any such events, he must have heard stories.

5 Langbein, *People*, pp. 321–22.

6 There was never a ramp at Monowitz, and the railway did not enter the camp; from 1942 on, standard procedure was that transports went to the "old Jew-ramp" at Oświęcim train station, or to a spur near Auschwitz I, and from 1944 to the ramp inside Birkenau; however, Fritz Kleinmann (Gärtner and Kleinmann, *Doch der Hund*, pp. 129–30) suggests that some transports were unloaded at or near Monowitz, presumably in open ground near the camp, and that men selected for Monowitz arrived with their luggage.

7 In Birkenau, two whole sections of the camp, known in camp slang as Kanada I and II, comprising thirty-six barrack blocks, were used for storage. Officially the sorting details were called "Aufräumungskommando" ("cleaning-up commando") but the unofficial name "Kanada Kommando" became so entrenched that the SS used it as well (Andrzej Strezelecki in Gutman and Berenbaum, *Anatomy*, pp. 250–52).

8 Although he makes no mention of it in his written memoir, Fritz says in his 1997 interview that he hoped Wocher would be able to find his mother, and gave him a letter for her.

9 There were twenty-three apartments in Im Werd 11; by 1941 and 1942 only twelve were still occupied (Lehmann directory house listings, Im Werd, 1938, 1941–42, WLO).

10 Ibid., 1942, WLO. It is not known whether Karl Novacek was related to Friedrich Novacek, who lived in the same building and was one of the friends who betrayed Gustav and Fritz in 1938 and 1939.

11 Transport list, Da 227, September 14, 1942, DOW. Transport Da 227 arrived at Minsk two days later, and as was usual the deportees were taken straight to Maly Trostinets and murdered (Alfred Gottwaldt, "Logik und Logistik von 1300 Eisenbahnkilometern" in Barton, *Ermordet*, p. 54). Bertha's daughter, Hilda, was married to Viktor Wilczek; Kurt Kleinmann's half-Jewish cousin and close friend Richard was their son.

Chapter 16: Far from Home

1 Gustav Kleinmann, letter to Olga Steyskal, January 3, 1944, DFK.

2 This restriction only applied in Monowitz; in the rest of Auschwitz, all categories of prisoner were eligible.

3 Langbein, *People*, p. 25; Fritz Kleinmann in Gärtner and Kleinmann, *Doch der Hund*, pp. 129–30.

4 Fritz Kleinmann in Gärtner and Kleinmann, *Doch der Hund*, pp. 129–30; Wagner, *IG Auschwitz*, pp. 101, 103; Levi, *Survival*, p. 32.

5 Fritz Kleinmann in Gärtner and Kleinmann, *Doch der Hund*, p. 132; Wagner, *IG Auschwitz*, p. 101.

6 Cesarani, *Final Solution*, p. 702. About 320,000 of Hungary's Jews had formerly been citizens of neighboring countries until Germany had carved off parts of them and given them to its Hungarian ally.

7 Ibid., p. 707. The Hungarian government resisted German calls for Jews to be deported. However, in August 1941 Hungarian border authorities took it upon themselves to deport some 18,000 Jews to German-occupied former Soviet territory. They were

murdered at Kamanets-Podolsk, Ukraine, along with about 8,000 local Jewish people, Cesarani, *Final Solution*, pp. 407–8; Rees, *Holocaust*, p. 292.

8 Danuta Czech, "Kalendarium der wichtigsten Ereignisse aus der Geschichte des KL Auschwitz" in Długoborski and Piper, *Auschwitz*, vol. 5, p. 201; Czech, *Auschwitz Chronicle*, p. 618.

9 Rees, *Holocaust*, pp. 381–82.

10 Danuta Czech, "Kalendarium der wichtigsten Ereignisse aus der Geschichte des KL Auschwitz" in Długoborski and Piper, *Auschwitz*, vol. 5, p. 203; Wachsmann, *KL*, pp. 457–61; Cesarani, *Final Solution*, pp. 707–11; Rees, *Holocaust*, pp. 381–85; Czech, *Auschwitz Chronicle*, p. 627.

11 Danuta Czech, "Kalendarium der wichtigsten Ereignisse aus der Geschichte des KL Auschwitz" in Długoborski and Piper, *Auschwitz*, vol. 5, p. 203.

12 Cesarani, *Final Solution*, p. 710.

13 Wachsmann, *KL*, pp. 460–61.

14 This appears to have happened around May 1944, as Gustav refers to it immediately after his description of the Hungarian Jews. In Fritz's memoir, he implies that it occurred before Christmas 1943, but the diary seems to rule this out.

15 Hospital admissions list, February–March 1944, pp. 288, 346, ABM. Gustav's illness isn't named in the record (which registers only name, number, dates, and either discharge, death, or "*nach Birkenau*"), and he doesn't refer to this episode in his diary, which jumps directly from October 1943 to May 1944.

16 Konstantin Simonov, quoted in Rees, *Holocaust*, p. 405. Other death camps in the region, such as Sobibór and Treblinka, had been decommissioned in October 1943, at the same time time as Maly Trostinets.

17 The other practical arguments were that aerial bombing was not precise enough to be effective. Hitting the gas chambers at Auschwitz, for instance, would have required such a magnitude of ordnance dropped over such a wide area that thousands of

prisoners in Birkenau would probably have been killed, without any certainty that the gas chambers would be hit. Bombing the rail network leading to the camps was similarly problematic. Railways were difficult to hit from high altitude, and wherever they were destroyed as part of the strategic campaign the Germans simply diverted traffic and usually had the tracks repaired and in service again within twenty-four hours. For overviews of the arguments on both sides, see Martin Gilbert, *Auschwitz and the Allies* (1981); David S. Wyman, "Why Auschwitz Wasn't Bombed" in Gutman and Berenbaum, *Anatomy*, pp. 569–87; Wachsmann, *KL*, pp. 494–96.

As for the question "Why didn't the Allies do something to halt the Holocaust?" this author's answer is that they did: they waged—and eventually won, at the cost of millions of Allied lives—an all-out war to destroy the state that was perpetrating it.

18 Air-raid precautions in Auschwitz had been discussed at a meeting of the camp command on November 9, 1943, including imposition of blackout, but nothing was apparently done until well into 1944 (Robert Jan van Pelt, *The Case for Auschwitz: Evidence from the Irving Trial* [2016], p. 328).

19 Some stricter Jews traded nonkosher foods for bread if they could, and there were Hasidic rabbis in Monowitz who refused all nonkosher food; they quickly starved to death (Wollheim Memorial oral histories: online at wollheim-memorial.de/en/juedische_reli gion_und_zionistische_aktivitaet_en; retrieved February 2, 2020).

20 "Even today the thought torments me," said Fritz many years later when he recalled his actions.

21 Fritz mentions this encounter in Gärtner and Kleinmann, *Doch der Hund* (p. 142) without identifying the young man more specifically. He appears to have been prisoner number 106468, who can be found in the Auschwitz III-Monowitz hospital record (ABM) but not in any other surviving Auschwitz records. This

serial number was one of a batch issued on March 6, 1943 to Jews deported from Germany (Czech, *Auschwitz Chronicle*, p. 347).

22 Wagner, *IG Auschwitz*, p. 108.

23 Fritz identifies them only by the names Jenö and Laczi. Surviving Auschwitz records show that two Jewish brothers arrived together on a transport from Hungary at about this time: Jenö and Alexander Berkovits (prisoner numbers A-4005 and A-4004; Monowitz hospital records and work register, ABM).

Chapter 17: Resistance and Betrayal

1 Without explanation, Fritz indicates that "Pawel" was also known as "Tadek." These were apparently false names. The real names of the Poles were Zenon Milaczewski (number 10433) and Jan Tomczyk (number 126261), although it isn't clear which was Szenek and which Pawel; the "Berliner" was apparently Polish-born Riwen Zurkowski (number unknown), who had presumably lived in Berlin (Czech, *Auschwitz Chronicle*, p. 619).

2 Fritz doesn't explain why Goslawski couldn't give the package directly to Peller at roll call. Possibly the construction workers were subjected to greater scrutiny when entering the factory enclosure. The date is variously given as May 4, 1944 (ibid.) or May 3 (Jan Tomczyk's prisoner record, ABM).

3 Monowitz commandant's office notification in Czech, *Auschwitz Chronicle*, p. 634.

4 Date unknown. Thirteen Poles were transferred to Buchenwald on June 1, 1944 (Czech, *Auschwitz Chronicle*, p. 638) and several transports of Poles went between August and December 1944 (Stein, *Buchenwald*, pp. 156, 166; Danuta Czech, "Kalendarium der wichtigsten Ereignisse aus der Geschichte des KL Auschwitz" in Długoborski and Piper, *Auschwitz*, vol. 5, p. 231).

5 The date of the execution is unclear. It may have been as late as December. The date of the death of Zenon Milaczewski (the real

name of one of them—see note 1 above) is given in the Monowitz hospital death book (ABM) as December 16, 1944.

6 Fritz states that two men were hanged, but according to Gustav
Herzog, there were three (Frankfurt trials statement, Abt 461 Nr
37638/84/15893, FTD).

7 Gilbert, *Auschwitz and the Allies*, p. 307. Gilbert states that the raid
began at 10:32 p.m., but this seems highly unlikely, as US bombing raids were normally performed in daylight. Czech (*Auschwitz
Chronicle*, p. 692) gives the time as "late afternoon."

8 Arie Hassenberg, quoted in Gilbert, *Auschwitz and the Allies*, p.
308.

9 Gilbert, *Auschwitz and the Allies*, p. 308; testimony of Siegfried
Pinkus, Nuremberg Military Tribunal: NI-10820: Nuremberg
Documents, quoted in Wollheim Memorial, wollheim-memorial.
de/en/luftangriffe_en (retrieved February 2, 2020).

10 Levi, *Survival*, pp. 137–38.

11 Czech, *Auschwitz Chronicle*, p. 722.

12 Gilbert, *Auschwitz and the Allies*, pp. 315ff.

13 Ibid., p. 326.

14 Prisoner number 68705, arrivals list, October 19, 1942, ABM;
Monowitz hospital records, ABM.

15 Prisoner number 68615, arrivals list, October 19, 1942, ABM.

16 Fritz doesn't identify the weapon as a Luger, but that's almost
certainly what it was. In Gärtner and Kleinmann, *Doch der Hund*
(p. 158) he describes it as a "0.8 mm pistol," which is clearly an
error. The model number of the military-issue Luger was P.08,
which might account for Fritz's error of memory. Luftwaffe units
were issued with the Luger well into the Second World War,
when higher-status army and SS units had switched to the
Walther P.38 (John Walter, *Luger: The Story of the World's Most
Famous Handgun* [2016], ch. 12).

17 In his memoir, Fritz mistakenly gives the date of this raid as
November 18. There was no air raid on that date. Altogether
there were four during 1944: August 20, September 13, and

December 18 and 26 (Gilbert, *Auschwitz and the Allies*, pp. 307–33).

18 Although many of the bombs fell on open ground, and a few on the surrounding camps, the December 18 raid did very heavy damage to several factory buildings (Gilbert, *Auschwitz and the Allies*, pp. 331–32).

19 Czech, *Auschwitz Chronicle*, p. 780.

20 Ibid., pp. 778–79.

21 Ibid., pp. 782–83.

22 Jósef Cyrankiewicz, January 17, 1945, quoted in Czech, *Auschwitz Chronicle*, p. 783.

23 Czech, *Auschwitz Chronicle*, pp. 785, 786–87.

24 Gustav Kleinmann's diary indicates units of 100, whereas other records specify 1,000 as the unit size (Czech, *Auschwitz Chronicle*, p. 786), and Fritz Kleinmann's memoir mentions three groups of about 3,000; the inference is that the units were organized hierarchically, in military style.

25 Gustav specifically identifies Moll. He was based at Birkenau, and no record has been found of his presence at Monowitz at this time. Possibly it was a flying visit to check on the evacuation.

26 On January 15, 1945, the total number of prisoners in Auschwitz III-Monowitz and its subcamps was 33,037 men and 2,044 women (Czech, *Auschwitz Chronicle*, p. 779).

Chapter 18: Death Train

1 Altogether, fifty prisoners were shot dead during the march (Czech, *Auschwitz Chronicle*, p. 786n).

2 Irena Strzelecka in Megargee, *USHMM Encyclopedia*, vol. 1A, pp. 243–44.

3 Four trains left Gleiwitz that day, carrying prisoners from several Auschwitz subcamps besides Monowitz. The Monowitz

prisoners were split between different trains, variously destined for the concentration camps of Sachsenhausen, Gross-Rosen, Mauthausen, and Buchenwald (Czech, *Auschwitz Chronicle*, p. 797).

4 Ibid., p. 791.

5 Moon phase data from www.timeanddate.com/moon/austria /amstetten?month=1&year=1945.

6 In his 1997 interview, Fritz says that he discarded his camp uniform after jumping, but in his written memoir he places it before. This seems more likely, since his uniform would be of value to the other prisoners to fend off the cold.

7 Eating regular soap would probably not have much effect (although the carbolic in use at the time might). Shaving soap, however, typically contains potassium hydroxide, which is highly toxic and produces severe gastrointestinal symptoms if ingested.

Chapter 19: Mauthausen

1 Mauthausen arrivals list, February 15, 1945, 1.1.26.1/1307365, ITS. Fritz jumped from the train on January 26, 1945 (per Gustav's diary, which agrees, give or take one day, with the record of the train's arrival at Mauthausen: AMM-Y-Karteikarten, PGM), and was entered on the records at Mauthausen on February 15 (Mauthausen transport list, AMM-Y-50-03-16, PGM)—eleven days later than by his own reckoning of his time in custody in St. Pölten.

2 Prisoner record card AMM-Y-Karteikarten, PGM; Mauthausen arrivals list, February 15, 1945, 1.1.26.1/1307365, ITS. Mauthausen received no documentation from Auschwitz about the transport of prisoners (for reasons revealed later in the chapter); hence Fritz's ability to pass himself off as Aryan. His tattoo was noted on his record as a distinguishing feature, but the number, which was meaningless here, wasn't taken down.

3 The liberation of Auschwitz attracted little attention in the press, despite Soviet attempts to publicize it. It was seen as a rerun of the previous summer's revelations about Majdanek, and was overshadowed by coverage of the Yalta conference of February 4–11. On February 16 (the day after Fritz Kleinmann entered Mauthausen) the first Western Allied serviceman to see inside Auschwitz after its liberation, Captain Robert M. Trimble of USAAF Eastern Command, was given a guided tour of Birkenau by Soviet officers (Lee Trimble with Jeremy Dronfield, *Beyond the Call* [2015], pp. 63ff.).

4 Prisoner record card AMM-Y-Karteikarten, PGM; Mauthausen arrivals list, February 15, 1945, 1.1.26.1/1307365, ITS.

5 Testimony of local priest Josef Radgeb, quoted in museum guide at mauthausen-memorial.org/en/Visit/Virtual-Tour#map||18 (February 2, 2020).

6 Czech, *Auschwitz Chronicle*, p. 797.

7 According to an account cited in Czech, *Auschwitz Chronicle*, p. 797n, the transport reached Nordhausen on January 28. This seems highly unlikely, since it had arrived at Mauthausen on January 26 and was kept there a whole day. Gustav Kleinmann gives February 4 as the date, which is much more likely.

8 The figure of 766 comes from Gustav's diary; the other figures are from Czech, *Auschwitz Chronicle*, p. 797n.

9 Mittelbau-Dora prisoner list, p. 434, 1.1.27.1/2536866, ITS.

10 Michael J. Neufeld in Megargee, *USHMM Encyclopedia*, vol. 1B, pp. 979–81.

11 According to Neufeld (ibid., p. 980), this extremely early start was practiced during the summer months, but Gustav Kleinmann's diary states that it was the case in February to March 1945.

12 A small camp had been established by this time at Woffleben (camp B-12) to save the journey time for workers from Ellrich (ibid., p. 981); however, Gustav and most of the other prisoners were not among those transferred here, and they continued having to make the journey to and from the site each day.

13 Langbein, *Against All Hope: Resistance in the Nazi Concentration Camps, 1938–1946,* trans. Harry Zohn (1994), pp. 374–75.

14 One theory is that the SS intended to use the volunteers as decoys, to draw enemy fire while the real SS made their escape (Evelyn Le Chêne, *Mauthausen: The History of a Death Camp* [1971], p. 155).

15 Fritz makes no mention of this episode in either his written memoir or his 1997 interview, and does not appear to have told his family about it. However, he did talk about it in a 1976 interview with fellow Austrian Auschwitz survivor and resistance member Hermann Langbein (Hermann Langbein, *Against All Hope,* p. 374).

16 Prisoner record card AMM-Y-Karteikarten, PGM; Gusen II transfer list, March 15, 1945, 1.1.26.1/1310718; Mauthausen transfer list, March 15, 1945, 1.1.26.1/1280723; Gusen II prisoner register, p. 82, 1.1.26.1/1307473, ITS. Langbein's sources (*Against All Hope,* p. 384) indicate that the plan to infiltrate the SS units occurred in "mid-March" 1945, but the episode must have been in early March, before Fritz's transfer to Gusen on March 15.

17 Robert G. Waite in Megargee, *USHMM Encyclopedia,* vol. 1B, pp. 919–21.

18 Gusen II transfer list, March 15, 1945, 1.1.26.1/1310718, ITS; Rudolf A. Haunschmied, Jan-Ruth Mills and Siegi Witzany-Durda, *St. Georgen-Gusen-Mauthausen: Concentration Camp Mauthausen Reconsidered* (2007), pp. 144, 172. In his memoir (Gärtner and Kleinmann, *Doch der Hund,* p. 170), which is very sketchy at this point, Fritz erroneously identifies the aircraft as the Me 109.

19 Haunschmied et al., *St. Georgen-Gusen-Mauthausen,* pp. 198, 210–11.

20 Quoted in Stanisław Dobosiewicz, *Mauthausen-Gusen: obóz zagłady* (1977), p. 384.

21 Dobosiewicz, *Mauthausen-Gusen,* p. 386. The only prisoners left behind were seven hundred invalids in the hospital, who were too sick to be moved.

22 Haunschmied et al., *St. Georgen-Gusen-Mauthausen*, pp. 134ff.

23 Ibid., pp. 219ff.

Chapter 20: The End of Days

1 Gustav gives no further details about Erich or his sources of food; most likely it came from civilians employed in armament production in the tunnel complex.

2 Michael J. Neufeld in Megargee, *USHMM Encyclopedia*, vol. 1B, p. 980.

3 Ibid., p. 970.

4 Ibid., p. 980.

5 Gustav names this place as Schneverdingen, north of Munster. This seems unlikely, since it would have necessitated immediately doubling back south to the ultimate destination. However, given the chaotic nature of concentration camp evacuations at this time, that would not have been out of the question.

6 David Cesarani, "A Brief History of Bergen-Belsen" in Suzanne Bardgett and David Cesarani, eds. *Belsen 1945: New Historical Perspectives* (2006), pp. 19–20.

7 Derrick Sington, *Belsen Uncovered* (1946), pp. 14, 18, 28; Raymond Phillips, *Trial of Josef Kramer and Forty-Four Others: The Belsen Trial* (1949), p. 195.

8 Langbein, *People*, p. 406.

9 Josef Rosenhaft, quoted in Sington, *Belsen Uncovered*, pp. 180–81; testimony of Harold le Druillenec in Phillips, *Trial*, p. 62.

10 Quoted in Sington, *Belsen Uncovered*, p. 182.

11 Celle was liberated by British forces on April 12, 1945.

12 Testimony of Captain Derrick A. Sington in Phillips, *Trial*, pp. 47–53; Sington, *Belsen Uncovered*, pp. 11–13.

13 Testimony of Captain Derrick A. Sington in Phillips, *Trial*, pp. 47, 51; Sington, *Belsen Uncovered*, pp. 14–15.

14 Sington, *Belsen Uncovered*, p. 16.

15 Ibid., p. 18.

16 Ibid., p. 187.

17 The original message has not survived, but Edith did receive it. It told her little other than that her father was alive and in block 83 of Bergen-Belsen (Samuel Barnet, letter to Sen. Leverett Saltonstall, June 1, 1945, War Refugee Board 0558 Folder 7: Requests for Specific Aid, FDR).

18 Molly Silva Jones in "Eyewitness Accounts" in Bardgett and Cesarani, *Belsen 1945*, p. 57.

19 Major Dick Williams, "The First Day in the Camp" in Bardgett and Cesarani, *Belsen 1945*, p. 30.

20 Ben Shepard, "The Medical Relief Effort at Belsen" in Bardgett and Cesarani, *Belsen 1945*, p. 39.

21 Molly Silva Jones in "Eyewitness Accounts" in Bardgett and Cesarani, *Belsen 1945*, p. 55.

22 Gerald Raperport in "Eyewitness Accounts" in Bardgett and Cesarani, *Belsen 1945*, pp. 58–59.

23 Haunschmied et al., *St. Georgen-Gusen-Mauthausen*, pp. 219ff; Dobosiewicz, *Mauthausen-Gusen*, p. 387.

24 It is unclear how many prisoners were herded into the Kellerbau tunnels, partly because of widely varying figures for the number of prisoners in the Mauthausen complex at the time. The total prisoner population of Mauthausen and Gusen has been given variously as 21,000 (Robert G. Waite in Megargee, *USHMM Encyclopedia*, vol. 1B, p. 902), 40,000 (Haunschmied et al., *St. Georgen-Gusen-Mauthausen*, p. 203), and 63,798 (Le Chêne, *Mauthausen*, pp. 169–70).

25 Fritz Kleinmann in Gärtner and Kleinmann, *Doch der Hund*, p. 171; Langbein, *Against All Hope*, p. 374; Le Chêne, *Mauthausen*, p. 165.

26 Krisztián Ungváry, "The Hungarian Theatre of War" in Karl-Heinz Frieser, *The Eastern Front, 1943–1944*, trans. Barry Smerin and Barbara Wilson (2017), pp. 950–54.

27 Le Chêne, *Mauthausen*, pp. 163–64.

28 George Dyer, quoted in Le Chêne, *Mauthausen*, p. 165.

29 Haunschmied et al., *St. Georgen-Gusen-Mauthausen*, p. 226.

30 Quoted in Langbein, *Against All Hope*, p. 82.

31 Gustav erroneously identifies this place as Ostenholz, a village to the southwest of Bergen-Belsen, well away from the route he and Josef Berger took.

Chapter 21: The Long Way Home

1 Samuel Barnet, letter to Sen. Leverett Saltonstall, June 1, 1945; William O'Dwyer, letter to Samuel Barnet, June 9, 1945, War Refugee Board 0558 Folder 7: Requests for Specific Aid, FDR.

2 Fritz does not identify the hospital, but it must have been the 107th EH, which established a facility at Regensburg on Apri 30, 1945 and remained there until May 20 (med-dept.com/unit-histories/107th-evacuation-hospital; retrieved February 2, 2020). No other American military hospital units have been identified in Regensburg at that time.

3 Fritz later researched the fates of fifty-five Jewish and non-Jewish children who had been playmates in the Karmelitermarkt before 1938 (Gärtner and Kleinmann, *Doch der Hund*, p. 179). Of the twenty-five Jews, five, including Fritz himself, survived the camps, and eight, including Kurt and Edith Kleinmann, either emigrated or hid. Twelve were murdered in the concentration camps. Of the thirty non-Jewish children, nineteen stayed in or around Vienna throughout the war, and eleven served in the Wehrmacht during the war; of these, only three survived.

4 Gustav had apparently taken up smoking since leaving Auschwitz.

5 Gustav identifies this man only as "G."

Epilogue

1 Naturalization records for Richard and Edith Patten, May 14, 1954: Connecticut District Court Naturalization Indexes, 1851–1992: NARA microfilm publication M2081.

2 On their testimony for the Frankfurt trials given in 1963, Gustav gave his religion as "Mosaic" (Jewish) and Fritz as "no religious affiliation" (Abt 461 Nr 37638/84/15904–6; Abt 461 Nr 37638/83/15661–3, FTD).

3 Statistics given in Gold, *Geschichte der Juden*, pp. 133–34.

4 Devin O. Pendas, *The Frankfurt Auschwitz Trial, 1963–1965* (2006), pp. 101–2.

5 Trials of Burger et al. and Mulka et al., Frankfurt, 1963; testimony of Gustav Kleinmann (Abt 461 Nr 37638/84/15904–6, FTD) and Fritz Kleinmann (Abt 461 Nr 37638/83/15661–3, FTD). Gustav was interviewed mainly about the death march and camp senior Jupp Windeck; Fritz's statement is mostly concerned with Windeck and SS-Sergeant Bernhard Rakers.

6 Along with his father's diary and commentaries by Reinhold Gärtner, Fritz's memoir was included in the book *Doch der Hund will nicht krepieren* (1995, 2012).

Index

Note: Page numbers followed by *n* indicate an endnote with relevant number.

Insights,
Interviews
& More . . .

A Message from Kurt Kleinmann

MORE THAN SEVENTY YEARS have already passed since the dreadful days described in this book. My family's story of survival, loss of life, and rescue encompasses all those connected to that period who experienced incarceration, who lost family members, or who were lucky enough to escape the Nazi regime. It is representative of all who suffered through those days and therefore needs to be never forgotten.

My father's and brother's experiences through six years in five different concentration camps are living testimony to the realities of the Holocaust. Their spirit of survival, the bond between father and son, their courage, as well as their luck, are beyond the comprehension of anyone now living, yet kept them alive throughout the entire ordeal.

My mother sensed the danger we were in as soon as Hitler annexed Austria. She helped and encouraged my eldest sister to escape to England in 1939. I lived under Nazi rule in Vienna for three years until my mother secured my passage to the United States in February 1941. That not only saved my life, but also brought me to the home of a loving family who treated me as if I were their own. My second sister was not so fortunate. Both she and my mother were eventually arrested and deported with thousands of other Jews to a death camp near Minsk. I have known for decades that they were killed there, and I have even visited the remote location where it took place, but was deeply moved, devastated in fact, to read in this book for the first time exactly how this event happened.

That my father and brother both survived their ordeal is miraculously detailed in this book. I was reunited with them when, drafted into military service in 1953, I returned to Vienna fifteen years after leaving. Over the subsequent years my wife, Dianne visited Vienna many times with our sons, who met their grandfather Gustav and uncle Fritz. There was a close family relationship that survived separation and the Holocaust, and has lasted ever since. Therefore I did not have trauma or animosity toward Vienna or Austria. That does not mean, though, that I can totally forgive or forget Austria's history. In 1966, my father and stepmother visited me and my sister in the United States. Besides us showing them the wonders of our new country, it also provided them the opportunity to meet my foster family in Massachusetts. That thankful and joyous union brought together those dear to me who were responsible for my existence and my survival.

The Boy Who Followed His Father into Auschwitz is a sensitive, vivid, yet moving and well-researched story of my family. It is difficult for me to describe my gratitude to Jeremy Dronfield for putting it together and writing this book. It is beautifully ▶

A Message from Kurt Kleinmann (*continued*)

written, interspersing the memories of myself and my sister with the story of my father and brother in the concentration camps. I am grateful and appreciative that my family's Holocaust story has been brought to the public's attention and will not be forgotten. ∿

Kurt Kleinmann, New Jersey, August 2018

Family Photographs

KLEINMANN FAMILY, APRIL 1938

This photograph was taken in Vienna in April 1938, one month after the Nazi annexation of Austria. Left to right: Herta, Gustav, Kurt, Fritz, Tini, Edith. The photo was Tini's idea; she had a foreboding sense that the family might not be together for much longer.

(Photograph courtesy of Peter Patten)

FRITZ KLEINMANN, C. 1938

Fritz at fifteen while a student at Vienna's technical high school, where he was training to be an upholsterer like his father.

(Photograph courtesy of Rebecca Hagler)

Gustav Kleinmann's Diary

The pages of Kleinmann's diary shown here, from November 1939, describe the beating to death of a rabbi, Fritz being whipped by the SS, and prisoners forced to stand in the roll-call square for hours as punishment. The page also describes how Gustav suffered a near-fatal bout of fever and only survived because of Fritz's care.

(Photograph courtesy of Visual History Archive)

Fritz and Gustav reunited

When Fritz and Gustav Kleinmann returned to Vienna in 1945, they were among only 26 survivors of the 1,035 Viennese Jewish men transported to Buchenwald concentration camp in October 1939.

(Photograph courtesy of Reinhold Gärtner)

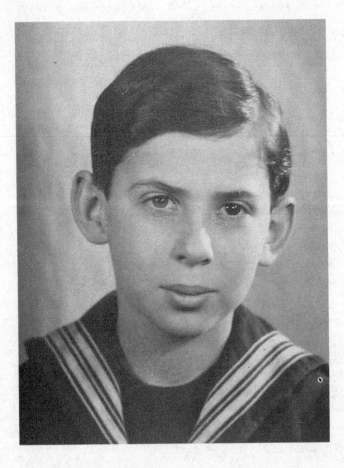

KURT KLEINMANN, C. 1938

Kurt Kleinmann at age eight in 1938. This photo was taken around the same time as the family photograph.

(Photograph courtesy of Rebecca Hagler)

Permission to reproduce the photographs has been secured from the following sources:

Peter Kleinmann: son of Fritz Kleinmann

Kurt Kleinmann: son of Gustav, brother of Fritz

Reinhold Gärtner: professor of political science and friend of Fritz Kleinmann

Rebecca Hagler: granddaughter of Edith Kleinmann

Peter Patten: son of Edith Kleinmann

Visual History Archive (VHA), University of Southern California Shoah Foundation

United States Holocaust Memorial Museum (USHMM), Washington, DC

Afterword by the author

WHAT YOU HAVE READ in these pages is a true story. Every person in it, every event, twist, and incredible coincidence, is taken from historical sources. One wishes that it were not true, that it had never occurred, so terrible and painful are some of its events. But it all happened, within memory of the still living.

I first discovered the story when I learned of the existence of Gustav Kleinmann's concentration camp diary. Alongside my work as a writer, I have a role as a publishing consultant. In February 2013 I was asked by Austrian academic Professor Reinhold Gärtner to help find a publisher for an English translation of Gustav's diary, coupled with a short memoir by his son Fritz and an academic commentary. I was immediately enthralled by their powerful story. But to my intense frustration, no publishers would take the book on. I had to acknowledge the fact that, as powerful and important as the diary is, it's extremely difficult to read. Gustav didn't write it as a literary work, but as a way of keeping a grip on his sanity. The diary is sketchy, making vague, cryptic references to people, places, and events. Even Holocaust historians would have to reach for their reference books to understand many parts of it.

I faced a dilemma. Studying the diary and Fritz's recollections, I recognized from the start that there was something truly special here. There are many Holocaust stories, but not like this one. The tale of Gustav and Fritz, father and son, contains elements of all the others but is unlike any of them. Very few Jews survived long enough in the Nazi concentration camps to witness them from the first mass arrests in the late 1930s all the way through to the Final Solution and eventual liberation. None, to my knowledge, went through the whole inferno together, father and son, from beginning to end, from living under Nazi occupation, to Buchenwald, to Auschwitz, to the death marches, and then on to Mauthausen, Mittelbau-Dora, Bergen-Belsen, and made it home again alive. Certainly none who left a written record. Luck and courage played a part, but what ultimately kept Gustav and Fritz living was their love and devotion to each other. "The boy is my greatest joy," Gustav wrote in Buchenwald. "We strengthen each other. We are one, inseparable."

I was profoundly moved by the test to which that bond had been subjected, when Gustav was transported to Auschwitz—a near-certain death sentence—and Fritz chose to cast aside his own safety in order to accompany him. I knew that this story *had* to be told far and wide. Having failed to publish the diary itself, I believed that the best way I could serve it would be to use my abilities as a researcher and writer, telling the story in a form that would be accessible to everyone.

I have brought the story to life with all my heart. Before setting down the first words, I spent a long period just considering how I would approach the writing—how it should flow, how descriptions, events, and ▶

characters should be handled. More than with any other book I've written, those elements had to be precisely right. In the end I settled on the example offered by the great works of survivor literature—especially *Night* by Elie Wiesel and the memoirs of Primo Levi. Their stark simplicity of form, coupled with an instinct for pure narrative, allows the events to speak directly to the reader.

When I finished, I had a book that reads like a novel. I'm a storyteller as much as a historian, and yet I hadn't needed to invent or embellish anything; even the fragments of dialogue are quoted or reconstructed from primary sources.

It wasn't always clear that it would turn out that way. The bedrock for my research was Gustav's diary, supplemented by written recollections and recorded interviews given by Fritz between 1979 and his death in 2009. But I needed more. There were large gaps that made the story untellable in the form it needed, and for a time I considered writing it as fiction, allowing myself the freedom to fill in the gaps with my imagination. But as I delved into the archives and the published accounts of other survivors—much of my research dedicated to deciphering the obscure references in Gustav's diary—the gaps slowly began to fill with factual information, coalescing into a vista rich and harrowing. My work on the book spanned two years of preliminary research and preparation followed by a year of full-time research and writing, bringing together a panorama of the Holocaust, from Vienna life in the 1930s to the flight of the refugees, from the functioning of the camps to the final liberation and homecoming. The documentary research, including survivor testimony, camp records, and other official documents, verified the story at every step of the way, even the most extraordinary and incredible passages. It revealed in full Gustav's indomitable strength and spirit of optimism: "... every day I say a prayer to myself," he wrote in his sixth year in the camps: "*Do not despair. Grit your teeth—the SS murderers must not beat you.*"

In shaping the book, it was important to me that I should tell not only Fritz's and Gustav's tale but those of the other family members, who are touched upon only lightly in the diary and Fritz's memoir. The heart of the story is a familial bond, and I needed to understand how that was forged. Starting out, I had no idea whether any members of the Kleinmann family were still alive. If they were, they might be willing to answer my questions. What was the Kleinmann family like? How did they live? What made them tick? How did it feel to be there in that time? Of those who had survived the Holocaust, Gustav and Fritz had passed on; that left only Fritz's siblings, Edith and Kurt. Eventually, through a young descendant of the family living in Chicago, I learned the sad news that Edith had only recently died. But Kurt was still alive and well in New York, a little way up the Hudson river from NYC. When I first heard his ▶

voice on the phone—with that unique, delightful blend of Massachusetts and Vienna—I felt my first real sense of *touching* the story.

It was the first of many long conversations, and the beginning of a friendship. We didn't get a chance to meet in person until a few years later, when Kurt came to London to help launch the book. For me it was like meeting a friend. I was deeply moved when Kurt said that I had delved so deeply into his family that it was like meeting a new brother. When I think who his real brother was—a boy and man whose strength and courage awe me—I am humbled.

On an emotional level, knowing Kurt personally made it harder to write this difficult and harrowing story. In my research, I uncovered a lot of information that had been unknown to the survivors— even things that Gustav and Fritz never knew. Most upsetting of all was my discovery of the fate of Tini and Herta Kleinmann. Naturally, Kurt knew that his mother and sister had been murdered, but not how. I still find this part of the story difficult to talk about. I wrote it knowing that it would bring pain to Kurt. When I emailed him the first draft of the book, I felt almost as if I were sending him a ticking bomb. He later told me that reading that particular episode devastated him, and he broke down. And yet he did not reproach me; he was glad to finally know the whole truth.

Given how many people this book has touched since it was first published, it's strange now to think how close it came to never being written. Conceiving it was one thing—finding a publisher for it was another matter entirely. As with the original attempt with the diary/memoir, my outline proposal was rejected by every British and American publisher it was sent to. "It's an amazing story," the rejections all said, "but there's no market for Holocaust books." I faced a difficult decision: Should I press on or abandon the project? I'm a professional writer, this is how I make my living—how could I write a book for no financial reward? By that time I had come to know Kurt, and felt I had incurred a moral obligation to him— and to the memory of the Shoah—to tell his family's story. So I pressed on.

It seemed that the book, once completed and modestly published by a small indie press in Chicago, would slip away into obscurity. Then, out of the blue, it was discovered by the visionary Dan Bunyard at Penguin Books in London. Against all the received wisdom of the industry, Dan encouraged me to hone and revise my original text, changing the title to *The Boy Who Followed His Father into Auschwitz.* Dan was convinced that this book could attract—and deserved to attract—a huge readership.

He was right. Of all the books I have written, *The Boy Who Followed His Father into Auschwitz* is the one I am most proud of. And it moves me that the book I value ▶

most is also the one that has been most widely read. At the time of writing this afterword, the story of the Kleinmann family has been translated into fourteen languages and has touched readers from Montevideo to Moscow. It is the most amazing story of courage and survival I have ever come across, and I feel privileged to have been entrusted with its telling. ᕁ

Jeremy Dronfield, England, February 2020